Climates in Miniature

CLIMATES IN MINIATURE

*A Study of
Micro-Climate and Environment*

by

T. BEDFORD FRANKLIN

GREENWOOD PRESS, PUBLISHERS
WESTPORT, CONNECTICUT

Wingate College Library

The Library of Congress has catalogued this publication as follows:

Library of Congress Cataloging in Publication Data

Franklin, Thomas Bedford.
   Climates in miniature.

   Reprint of the 1955 ed.
   1. Microclimatology. I. Title.
[QC982.7.F7 1972]   551.6'6   79-138234
ISBN 0-8371-5591-6

All rights reserved

Originally published in 1955
by Philosophical Library, New York

Reprinted with the permission
of Philosophical Library

First Greenwood Reprinting 1972

Library of Congress Catalogue Card Number 79-138234

ISBN 0-8371-5591-6

Printed in the United States of America

Contents

I.	WHAT A MICRO-CLIMATE IS	page	11
II.	WARMTH AND HIBERNATION		15
III.	CRANBERRY MARSHES		25
IV.	SOILS		34
V.	TAKING EARTH TEMPERATURES		42
VI.	WHAT HAPPENS TO THE RAIN?		50
VII.	NATURE'S WAY		58
VIII.	FROST		67
IX.	AIR TEMPERATURES		77
X.	HUMIDITY AND DEW		86
XI.	WIND AND SHELTER		94
XII.	LIGHT AND SHADE		102
XIII.	CLOCHES AND FRAMES		111
XIV.	MICRO-CLIMATES ON THE FARM		120
XV.	MICRO-CLIMATES IN THE HOME		129
	INDEX		135

Illustrations

PLATES

1. Spring flowers in a coppiced wood   *facing page* 36
2. Frozen spray saves fruit tree from frost damage   37
   (*Photograph by courtesy of the East Malling Research Station*)
3. A frost hollow   44
4. Fruit ripens well on a south wall under a wide coping   45
5. Dewdrops on twigs and berries after a calm clear night   80
6. A dew pond on the South Downs   81
   (*Photograph by Edward Reeves*)
7. On windy sites branches grow down wind   96
8. Even in winter trees give shelter from radiation   97

I
What A Micro-Climate Is

The study of micro-climates and environments is one of the newest forms of scientific research. In these ultra-scientific days, when the professional scientist has already produced jet propulsion, space rockets and atomic bombs, it is good to know that there is still a field where the amateur scientist can be of use with a minimum of equipment and expense.

As its name implies the study of micro-climates is the study of the climate of a very small area, and whether our interests are in the animal, bird, and insect life round our homes, or just in pure gardening there is scope in micro-climates for us all. There is nothing difficult about it, and if we are keen on gardening it means simply that we no longer think of our garden, however small it may be, as being the same all over, but we realize that even in it there are differences of soil, sunshine, temperature and humidity. Once we have realized this and done some simple experiments to prove it, we no longer think of it as a small garden, but as a small research station, where the beds are experimental plots, and where we can use our own ingenuity to modify and improve the natural conditions of soil and climate.

During the last thirty years I have done a very large number of simple experiments in and around my garden, some of which are described in the following chapters, and I still have many more I hope to do in the future. Those who have the opportunity of using a wider field than their own garden will find here experiments on the climate of rabbits' burrows and badgers' sets, on the hibernating refuges of hedgehog and dormouse, on the coming of spring in hedgerow and coppice; while the pure gardener will find records of the behaviour of different soils, of earth

WHAT A MICRO-CLIMATE IS

temperatures, of what happens to the rain, of the value of cloches and frames, and the climate of the laurel bush where my pair of blackbirds roost.

These and many more go to show that although the title contains the word climate this is not just another book about the weather. In it you will not find synoptic charts with depressions over Iceland and anticyclones over the Azores, for in our garden we take the weather as it comes and do what we can to modify it to our own purposes. Our most ambitious forecasts are only to decide whether it is likely to freeze to-night and if so what we are going to do about it. Nor will you find descriptions of the battery of instruments that every meteorological station is bound to keep or of the Stevenson screen which houses so many of them. For weather forecasting it is essential that the instruments should be of standard pattern and that they should be exposed in the same way at all stations; but with us our instruments will be as simple and inexpensive as possible so long as they will do the work required of them, and they will be exposed wherever animals and birds live or plants grow. The subjects of our experiments are so various that our instruments may be recording climates and environments several feet underground or in the branches of a tree.

Like the ancients we shall realize that animals, birds and insects use the micro-climates of their environment to their own advantage and can teach us a great deal, for though man's interest in local climates has been casual in the past, the birds and beasts have always realized that they had a considerable influence on the comfort of their lives. Fifty years ago I learnt my first lessons on earth temperatures from a rabbit's air-conditioned burrow, and from the way a mole adapted himself to winter frost by making his runs under long grass where the ground was less likely to freeze. In summer when the nights are warm cattle lie in the long grass where it is cool, but in spring and autumn when the nights are cold they seek out the warmer bare earth to lie on and are particularly fond of a macadam road. Motorists should know this well and drive very cautiously at night on unfenced roads across commons where cattle graze.

WHAT A MICRO-CLIMATE IS

The plover chooses a light sandy soil for her nest if she can find it because this soil warms up quickly in the spring, partridges roost on high ground, where the cold air drains away, and not in the hollows where it collects. A well-made form of a rabbit or hare is several degrees warmer than the open ground on a cold night. Insects are particularly susceptible to changes of humidity; the wood louse lying underneath a slate too hot to touch in full sunshine is quite happy as long as the ground beneath the slate is damp and the humidity high, but put under the same slate on dry sandy soil it will die quickly if it cannot escape and seek shelter elsewhere. Locusts have to suffer a daily loss of water by evaporation and would die if they could not replenish their store of water. So in the deserts they work in the early hours of the morning, drinking the dew which forms for a very short time on the drought-resisting vegetation which during the day is bone dry.

Many wild flowers appear to be very tolerant of extremes of temperature and can be found in flower when winter is still at its coldest, but in general it will be found that their choice of micro-environment and not their innate hardiness is the secret of their success. Removed from their chosen habitat and transplanted into our gardens they show their dislike of icy blast and frozen ground just as our ordinary plants do. But in their chosen environment of the woods, protected from frost by the canopy of trees and deep leaf mould soil, and yet open to the mild winds when they arrive, nature achieves a growing temperature many weeks before it arrives in the open.

In contrast dahlias are extremely intolerant of frost and when planted in a bed under a south-facing wall survive the first frost or two close to the wall because of the extra warmth and shelter from radiation provided by the wall. Outside a distance away from the wall which depends mainly on its height they will be cut down by the first frost, and we can run a garden line between the dead and the living which will be parallel to the wall, and thermometers on both sides of this line will confirm that this is the frost line.

The Benedictine and Cistercian monks solved the problems

WHAT A MICRO-CLIMATE IS

of micro-climates in their walled gardens with a thoroughness that was typical of all their farming and gardening operations, and by doing so produced fine crops of fruit and vegetables. In Scotland, in a not naturally favourable climate, the fruit from the orchard at Dryburgh and the apricots of Pluscarden were famous, while in the warmer climate of southern England they grew grapes in the open and made wine from them. They estimated the effects of the walls of their gardens on the borders underneath them by comparing the time of germination of seeds sown in rows parallel to the walls, and rightly concluded that both the height of the wall and the material of which it was made came into the question.

And since the germination of seeds depends not only on the warmth of the air above but also of the soil beneath, the study of underground temperatures will naturally arise, and we shall learn something of that great reservoir of heat which is just below the surface and which is filled by the sun every day it shines and is emptied every night to prevent the surface falling too low after sunset. But for this reservoir most of our plants would be scorched by day and frozen by night beyond recovery.

To anyone who has a piece of ground, however small, the study of micro-climates and environments is a rewarding hobby. The field of research is all around us and the choice of experiments is very wide, and though the amount of necessary equipment and expense can be kept quite small the usefulness of our results may be out of all proportion to the time and money we are able to expend on the work.

As the ordinary reader is more accustomed to using Fahrenheit thermometers and inches in his calculations, my temperature records are all in F.° and my measurements in feet and inches, rather than in the C.° and centimetres of the scientist.

II

Warmth and Hibernation

Every winter as soon as the keen frosts begin a wild rabbit visits my town garden and reopens the burrow made under my compost heap. Modern compost made properly with the aid of a chemical organic activator generates a considerable amount of heat, and a centrally heated burrow must be a desirable luxury in comparison with the somewhat shallow burrow in the railway cutting which may be all right as a summer home but is too cold to be comfortable in mid-winter. For rabbits seem to be very susceptible to changes of temperature in spite of their excellent fur coat.

Fifty years ago in Northamptonshire other rabbits gave me my first knowledge and experiments on micro-climates, and it is a pleasing coincidence that to-day in Edinburgh another rabbit should confirm the conclusions come to at that earlier date. These conclusions, confirmed and extended by soil scientists, provided the means of saving a valuable fruit crop in the United States, as we shall see in the next chapter.

Years of experience of ferreting rabbits at all seasons of the year, sometimes to trap them as a pest, at others to shoot them for sport, had taught me that rabbits lived at different levels in their burrows at different seasons. In spring and summer the does use the shallow burrows to bring up their families in some stop run. Badgers know this and can locate nests of young rabbits from the surface and dig vertically down to them, and it is no great task to dig out a ferret that has laid up with a doe at her nest. But in late autumn and winter, when the breeding season is over, the rabbits go down to the deeper levels which are much warmer than the shallow ones and

WARMTH AND HIBERNATION

quite immune from frost, and it may take several hours of hard digging to recover a line ferret which has laid up at the end of its line in one of the deep levels of a warren. In winter, when ferrets are used for bolting rabbits for shooting, it is noticeable if the burrows are by a hedgerow and deep ditch how often the rabbits bolt from the lowest bolt holes in the ditch bank and not from the surface holes.

To get some records of temperatures in rabbit burrows I decided to instal maximum and minimum thermometers in them and keep a record of the extreme temperatures of each month at several levels. This would give some idea of the choice of temperature available to the rabbits by moving up or down a flat or two at any time of year. I found that thermometers could not be put into burrows in actual use by the rabbits as they invariably scratched them out and broke them, but I used a deserted burrow of the same warren where there were bolt holes into a ditch and there I installed my thermometers at arm's length up the holes and at approximately 4, 18 and 30 inches below ground level. I also made a hole the length of my arm in the occupied burrow area and put a control thermometer at 4 inches below ground level in this hole.

It was fortunate that I did this as it was soon apparent that the maximum readings of the control thermometer were always higher than those of the corresponding thermometer in the deserted burrow. The only difference between the inhabited and deserted burrows was that the ground all round the used burrow was covered with sandy spoil from the many holes and very little grass was showing, while the grass had grown again over the deserted burrow and was the same as the rest of the grass field. It seemed probable that the sandy surface over the used burrow got very hot in the sun and conducted this heat down to the 4-inch depth thermometer beneath. To test this I dug and spread a good deal of sandy soil over the ground above the deserted burrow to make it look as much like the used burrow as possible, and in a couple of days my thermometers were both registering the same temperatures at the 4-inch depth.

The accompanying graph—Fig. 1—gives the extremes of

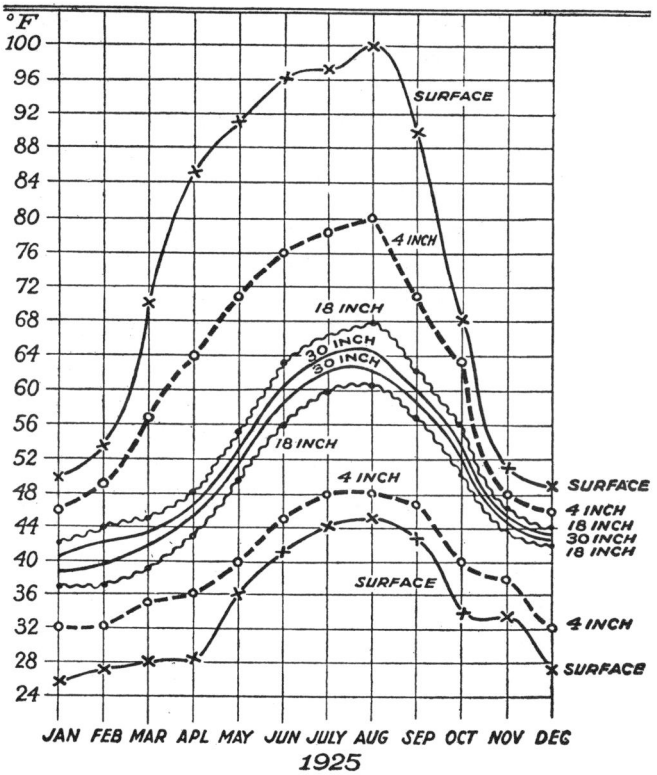

FIGURE 1

temperature at the surface of the soil and in burrows at 4, 18 and 30 inches depth below ground, and for comparison Table 1

TABLE 1
Mean Soil Temperatures at various depths in a gravelly soil under turf at Oxford

Depth	Jan.	Feb.	Mar	Apr.	May	June	July	Aug.	Sep.	Oct.	Nov.	Dec.
4 inch	36	36	42	52	60	66	70	70	65	55	43	37
6 inch	39	39	42	48	55	62	66	64	59	52	45	41
1ft. 6ins.	40	40	42	47	53	59	63	63	59	53	47	43
3ft. 6ins.	43	42	43	46	51	56	60	61	59	55	50	46
10 ft.	49	48	47	47	48	50	53	55	56	55	54	51

WARMTH AND HIBERNATION

gives the mean temperatures at somewhat different depths in gravelly soil under a turf surface at Oxford from records taken over ten years. By going down to the 18 or 30 inch level the rabbits could avoid the extremes of the 4-inch level which was too hot in the summer and too cold in the winter. Most of the young ones in this warren appeared to be born at the 18-inch level where temperature ranged from 50° to 68° from May to October. An equable micro-climate without any great extremes of heat or cold seems to be the choice of rabbits if they can get it; but in certain places, where the surface soil was of little depth and the subsoil was nearly solid rock, burrows are found which at most can only be a few inches below the surface. Here at the height of summer the temperature and probably the humidity become intolerable and the rabbit leaves his burrow to lie out in the long grass in the open where the breeze and the shade are more to his liking.

I was able to find two small burrows close together in very shallow sandy soil, one in full sun all day and the other sheltered by trees from the morning and midday sun and only in sunshine for the late afternoon and evening. Handy to both sets of burrows there was a large patch of uncultivated grass full of thorn bushes and saplings and an ideal patch for the rabbits to lie out on. Yet whenever I walked through the patch in the heat of the day and disturbed those rabbits that lay out there they nearly all bolted to the burrows in full sun and not to the burrows under the trees in the shade. This seemed to show that shallow burrows are unpleasantly hot for rabbits when the sun is strong in the summer.

A black rabbit lived in the shallow sunny burrows and was invariably to be found lying out in the daytime. But I was surprised to find that it continued this practice right through the autumn and that it lay in its form several times when hoar frost was on the grass. Perhaps its black coat was warmer in the sun than the brown coats of the other rabbits and the temperature of its form might be higher than that of the open air and quite possibly no colder than the shallow burrow it had to go to otherwise.

WARMTH AND HIBERNATION

Stoke Park Wood in Northamptonshire had a couple of badgers in residence and their set housed not only the badgers but a fox and vixen and many rabbits. Certainly they all had different front doors, and the rabbits were safe when at home in burrows too small for badger or fox; but it must have been rather exciting work for the rabbits living so close to the badgers who dug out and ate their young, and to the fox who would snap up any unsuspecting rabbit, young or old, that he came across on his way to and from his earth. This was under a tree root on the top of a bank and six feet above the entrance to the badgers' set at the foot of the bank. The bank was probably the edge of an old marl pit for it ran in almost a complete circle with a gap where a horse and cart could enter and load up the marl. In the middle of the circle was a mound, made probably of the last load of marl which had never been removed, and this mound in the middle of a natural amphitheatre was the common playground of the young rabbits by day, the fox cubs at dusk, and the badger cubs after dark. There must have been a fine mixture of scents on the mound, but the young seemed to have no inhibitions, though the adult rabbits never went there, the vixen lay at the mouth of her earth and watched her cubs below at play, and as far as one could see in the dark, only the badger cubs used it at night.

It was much more usual for the vixen to play with her cubs, and at another earth not far away I watched a game that I had the luck to initiate. The earth was under a tree stump at the foot of a mound on which cattle would throw earth over themselves with their hooves when troubled with flies. One afternoon, sitting on the mound, I made a set of mud pies from this earth and left them on the top of the mound when I came away. On the morrow I was surprised to find my mud pies at the bottom of the mound, so I replaced them on the top and that evening lay hidden to solve the riddle. About dusk three fox cubs emerged from the earth followed by the vixen. The cubs dashed up the mound and rolled down my mud pies, tumbling and somersaulting after them in great delight. As their mouths were too small to carry the mud pies up again, the vixen replaced them and the game was repeated again and again until the mother was tired of it and would carry no more. And so I always found my mud pies at the bottom of the mound in the morning.

By digging away part of the mound of earth in front of the entrance to the badgers' set I could push a six-foot bamboo cane with thermometer attached into the set. At this depth the temperature varied little throughout the year, being about 45° all winter and rising to about 55° in August. That the badgers lived at that depth was probable as they collected fresh bedding and took it in by this entrance and also pushed out the soiled and discarded bedding by the same entrance. There was another entrance on the top of the bank but this seemed to be deserted, probably because it was too near the fox's earth.

The badger is capable of doing without food for several days at a time if necessary, but does not hibernate. Warm-blooded animals which hibernate have to relinquish their temperature regulation and become cold blooded, so that in hibernation the temperature of their extremities becomes as cold as their surroundings, their circulation is very sluggish, their heart-beats very slow, and their respiration so slow that it is difficult to detect at all. The badgers, living as they do in a temperature always between 45° and 55°, would probably find all this impossible, for though the modern idea that the fat deposits that hibernating animals put on in the autumn provide the stimulus to hibernate rather than the colder temperature and shortage of food, yet it is undoubted that a rise in temperature often arouses hibernators and the times of hibernation and its duration vary from season to season according to the weather.

At my preparatory school we were allowed to keep pets in a pets' room which was centrally heated in the winter, and my pet was a dormouse. In the carpenter's shop I made him a good sized cage consisting of an exercise ground and a miniature tree to climb to get to his nest box. So long as the cage was kept in the pets' room at a temperature between 45° and 55° most of the winter, the dormouse woke every evening for his supper of nuts, haws, or pieces of apple and never hibernated. The evening meal over, he had his exercise, which consisted of running up the tree, going into his sleeping box, sliding down a boxed-in ramp from

WARMTH AND HIBERNATION

the sleeping box to the floor and peering out through a hole leading from the floor to the cage. From there he would watch to see if he had an audience, and if the audience laughed or clapped their hands he would repeat the performance indefinitely. From the look in his eyes as he peeped through the hole he must have had a sense of humour and enjoyed our admiration of his agility. The next year, as I had been told it was cruel not to let him hibernate, I put the cage in an unfired room with a temperature about 40° most of the winter, and he slept without waking up for food from the end of September to March and was very thin when he woke.

In our garden at home in Northamptonshire we had a tame hedgehog and there we kept more elaborate temperature records. All summer piggy came out about six o'clock in the evening for his supper of bread and milk of which he was extremely fond. We only had to whistle or call 'piggy' and he would come hurrying along from some garden bed where he was busy feeding. Being more or less a pet and living in a walled garden where none of his usual enemies ever came, he soon forgot his habit of feeding only at night and became an afternoon and evening feeder as well. On the approach of winter he got his hibernating quarters ready in a hole under a large tree stump in a shrubbery. There he made his winter nest of leaves and moss and on November 15th he blocked up the hole with a mass of leaves and went to sleep.

We bored a hole through the wood above his nest and fitted it with a thermometer packed tight with clay while he was getting his bed ready and as soon as he started hibernating we kept records of the temperatures in his refuge during the winter. The minimum temperature he had to endure that winter was 36° and was after quite a long spell of hard frosts down to 18° on the grass, but the shelter of the evergreen shrubbery, the tree stump and his bedding kept out the frost. In January there was a long spell of depressions with westerly wind and warm rain interspersed with cold north-westerly winds and fine weather, but we found that it took at least two days for the extremes of warmth and cold to penetrate into his fastness so that unless the spells of

WARMTH AND HIBERNATION

warmth lasted more than two days they had no effect on piggy. But an eight-day spell of south-westerly wind and a temperature of 52° woke him and he emerged for a day or two but hurried back as soon as the inevitable north-west wind and snow showers returned.

Our second hedgehog was in our garden in Derbyshire and was not nearly so tame as the other, as with a large wild garden to wander in he had no need for extra food and refused to be fed. It was really only by accident that we discovered his hibernating place when we saw him carrying grass and leaves into an old wasps' nest in a bank covered with moss and long grass and about six inches below the level of the ground. We planted a six-inch depth thermometer by the side of his nest to give us some idea of his hibernating temperatures. During the mild autumn he was still about and did not settle in until December 2nd when slight snow fell and there was a keen frost; after that he closed the hole of the wasps' nest with moss and apparently slept until the upward surge of earth temperatures in April. We kept the hole of his nest under observation daily and never found it unblocked so probably he did sleep for nearly five months. The lowest temperature recorded at the six-inch depth during the winter was 35° and he reappeared after the six-inch temperature had stood at 50° for several days in a mild spell at the end of April.

Both hedgehogs were clever enough to choose a hibernating retreat that was frost-proof in the years we had them under observation, and though temperature may not be the prime cause of the urge to hibernate it does seem as if a temperature of about 50° for a day or two wakes them up either temporarily or permanently if the weather remains mild. Our second hedgehog owed his comfortable winter largely to having a good depth of snow cover over his retreat when the hardest frosts came; the following year he did not use the wasps' nest again, and perhaps wisely, for had he done so he would probably have been frozen to death as hard frosts without snow cover penetrated about eight inches into the ground and the temperature at the six-inch depth fell to 30°.

WARMTH AND HIBERNATION

There is a firm belief in Derbyshire and many other counties that you can foretell whether the winter is to be cold or mild by the choice the hedgehog makes for his retreat for hibernation. If he hides deeply underground or under a tree stump it will be a hard winter, while if he merely rolls himself up in a ball of grass and leaves in a hedgerow it will be mild. I am afraid this belief will not stand up to test, for after the hard winter when our hedgehog deserted his wasps' nest, many others of his kind hibernated in hedges and stone walls and the following spring my neighbour and his dog found several dead hedgehogs in places that obviously gave little protection against hard frost when there was no snow cover.

Birds are quite clever at finding warm roosts at night. I have a hen blackbird in my garden which is always with me when I am digging; she often sits on my shoulder watching for worms to be turned up, when she drops down and captures them and then returns to her perch. She seems to think that my job in life is to provide worms for her all the year round and does not take kindly to the frosty weather when garden work cannot go on. She roosts in a laurel that has always been trimmed into a solid dome-shaped bush so that it gives admirable shelter from rain and wind. It keeps out a lot of frost too as the falling cold air shoots off it to the ground, and I have found by keeping a minimum thermometer near her perch that on nights of hard frost it can be as much as 8° warmer than the open, and on many nights of cold wind it is about 4° warmer than the air outside. The hen blackbird evidently knows this is a good roost and guards it, as she guards the rest of her territory, from all intruders with a most determined ferocity.

She is, I think, a more intelligent bird than most as she has been busy this last week feeding her three young ones in the garden and was obviously tired of the job. So when she found some mutton fat off the roasting tin, that I had put on the compost heap, she divided it into three pieces, set a youngster in front of each piece, and then superintended the drill of young birds feeding themselves. It was a delightful performance and she puffed herself out with pleasure when she saw she had an

WARMTH AND HIBERNATION

audience. In the summer she combines sunbathing and a Turkish bath under a spare cloche that I leave for her on the corner of the lawn. She lies under it flat on the grass with wings and tail spread and beak wide open as if she was panting for air. I have to leave both ends of the cloche open as my neighbour's cat likes sunbathing too and my blackbird needs both front and back door open when the cat is about.

III

Cranberry Marshes

Before beginning any serious study of a micro-climate it is essential to choose out the factors on which the whole research is to hinge, and then consider what effect all the other factors may have on the chosen ones. This may take a little time, and may involve a quite considerable search through any previous work done on the same subject, but it is well worth while for nothing is more galling than to find, after spending a good deal of time on some problem, that the work has been done before, and that what we thought was a piece of original research is only confirmation of work done by someone else. To make the importance of this preliminary study clear, I cannot do better than give in some detail the factors involved in the study of the micro-climate of a cranberry marsh in Wisconsin, U.S.A., carried out by the Weather Bureau of the United States Department of Agriculture in 1907.

At that time the cultivation of cranberries was confined to three states—Massachusetts, New Jersey, and Wisconsin. Of these the Cape Cod marshes in Massachusetts were famous for their high state of cultivation, their drainage, and their lack of weeds. The vines there were only about five inches high and were so covered with berries that the vines were hardly noticeable. In contrast the Wisconsin marshes often had a rank growth of weeds through which the vines struggled up to eight or nine inches in height, and even when they were bearing freely it was the vines and not the berries that were noticeable. In consequence the yield per acre at Cape Cod was double that at Wisconsin and the total yield of the Massachusetts crop was 300,000 barrels against Wisconsin's 75,000 barrels.

CRANBERRY MARSHES

The superiority of the Cape Cod marshes was largely man made, for though the temperatures in the summer in Massachusetts and southern Wisconsin were practically the same, yet the minimum temperatures in the Cape Cod marshes averaged about 5° higher than in the marshes at Mather in Wisconsin. In many respects all the cranberry grounds were much the same; on calm clear nights the marshes filled with cold air which had drained off the surrounding hills to the lower levels where the cranberries grew, and the peat by its capillarity brought moisture to the surface and the cold produced by the evaporation of this moisture was considerable, especially if the drainage system was poor. As flooding the marshes with water from reservoirs was the only means of preventing damage to the berries by frost, the growers had to keep a certain amount of water in the ditches to expedite the flow of the water as otherwise the flooding took about six hours and was often too late to prevent frost damage. Thus a frost precaution kept the surface of the peat wet and by increasing the evaporation and cooling of the ground made the chance of a frost more likely.

In all the marshes the natural vegetation was generally dense and effectively screened the soil from the sun, so that it was heated very little by day and had only a small supply of heat stored up at night to counterbalance the outgoing radiation. Also leaves and grasses were excellent radiators and lost their heat very quickly; in consequence the surface of a bog covered with a dense growth of weeds had a poor supply of heat and lost it very rapidly, so that the temperature of the air above the bog fell to a low level on every calm clear night. The only advantage against frost that the marshes enjoyed was the large amount of moisture in the air, so that a little of the radiation from the ground and vegetation was absorbed and reflected back by the moist air and so not completely wasted. Also moist air, when cooled below its dew point, had to give up some of its moisture as dew, and in this condensation of moisture or hoar frost, if the dew point was below freezing point, a considerable amount of latent heat was liberated which helped to slow down the fall of temperature. But a large amount of moisture in the air greatly

CRANBERRY MARSHES

increased the tendency for fog which often lay over the marshes in the morning and did not clear till after midday, thus cutting off much of the sunshine and making a frost on the following night more likely. Frost in the Wisconsin marshes was undoubtedly the worst enemy of the cranberry. In an average year the buds swell about May 20th and flower from June 10th to July 1st and during this time the fruit is setting. The new terminal bud is formed by about August 5th and is very tender until about September 1st. As the fruit ripens it becomes more immune to frost and can only be damaged by a severe frost when fully ripe. The green berry can stand a temperature of 28° and a fully ripe berry of 24° for a short time. The most favourable weather conditions for hard frost in Wisconsin were an overcast day with a fresh wind so that the store of heat in the soil was small and mostly lost by evaporation, followed by a calm clear night with a rising barometer, so that radiation was active and the cold air settled to ground level, and low humidity which allowed the radiation from the ground to pass away to outer space without absorption by the water vapour in the air.

All these conditions were present on 7th August 1904 so that an abnormally low temperature was likely on the morning of August 8th. This was realized and the minimum temperature among the vines at Cranmoor was 26°, or about 20° below normal for that time of year. This was disastrous to the cranberries and as a result of the freeze on that day the crop of that year was reduced 40 per cent, and 25 per cent of the terminal buds for the next year's crop were killed. Obviously something had to be done for Wisconsin, and it was this frost which decided that the Weather Bureau should make an investigation on the possibility of frost protection.

Many of the factors involved were natural phenomena outside the control of man—the draining of the cold air from the hills to the marsh, the moist air, dew, and fogs—but man could at least do something to check evaporation and the waste of the sun's heat by useless weeds shading the ground and throwing away much needed warmth by their radiation. There was nothing against weeding the plots as they did at Cape Cod, but the draining and sanding of the marshes practised there could not be followed so easily, because on Cape Cod, near the sea, there was not the same risk of frost as at Wisconsin and the drains were not kept half-full of water to ensure a quick flooding of the marsh when there was a warning of frost.

On fairly dry peat at Cape Cod a two-inch covering of coarse sand was most effective; its lack of capillarity reduced the moisture brought to the surface and evaporated; its low specific heat made it warm up quickly in the sun and pass this heat down to the layers of peat below; its absorption of the sun's radiation was good, but it gave back its heat at night slowly and so modified the night temperature by warming the air above it during most of the night. But how would sand act on the wet peat at Wisconsin; obviously only an extended trial over a season would show. But at least the problem had been simplified because the chosen factors on which the research was to hinge were now reduced to weeding and sanding, all the other factors apparently being outside the control of man.

A start was made in September 1906 to find the effect of weeds and vegetation by comparing two plots, one of bare peat with no vegetation at all and the other plot with complete marsh vegetation. These two plots were close together and got the same amount of sun, but as the plots were very small the air temperatures over them were not reliable as the warm air over the peat drifted on to the marsh and the cooler air of the marsh drifted back on to the peat. But the underground temperatures gave a very decided answer in favour of the bare peat. Figure 2 shows that the range of temperature at the three-inch depth in the soil was practically negligible on the plot covered with complete marsh vegetation and the shaded area between the curves for maximum and minimum temperatures had hardly any area. As this shaded area represents heat usefully stored ready to be returned at night, a hard frost was almost certain on the marsh plot any night when conditions were favourable to frost.

In comparison with this the bare peat gave a large shaded area so that under it a large amount of heat was stored ready to re-

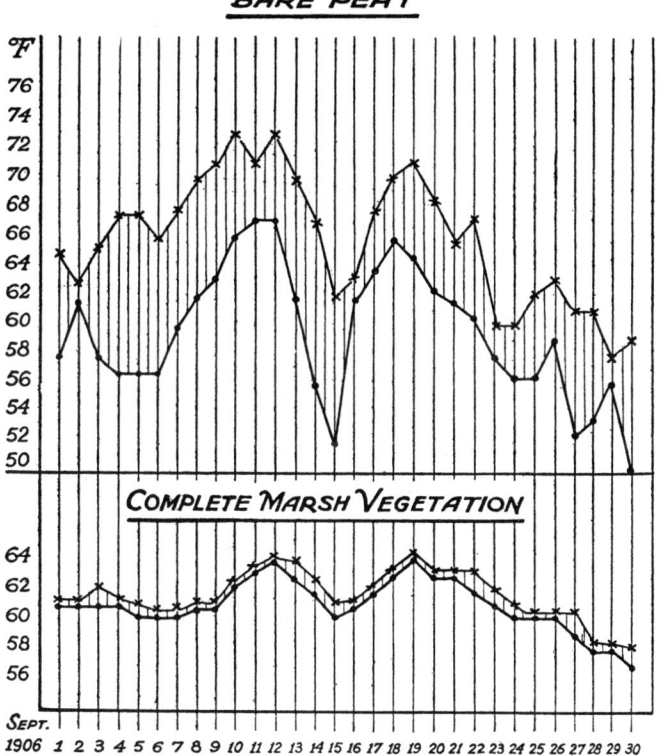

FIGURE 2

turn to the surface and help to ward off the frost, and in consequence the mean minimum surface temperature for the month was 6° higher on the bare peat than on the complete marsh vegetation plot. There was no doubt that useless vegetation in the form of weeds, moss and grass were very frost producing, and that on a well-weeded plot the cranberry vines tended to be dwarfed with less leaf and more fruit. On a weedy plot most of the sun's radiation was used in producing rank and useless growth, while on the weeded plot it was used to produce more fruit and to preserve it from frost. The answer was so definite that this experiment was not done again in 1907 when the ques-

CRANBERRY MARSHES

tion of the advantage of sanding the plots was tackled throughout that year.

In May 1907 several more stations were installed and fitted with thermometers; five of these are of chief interest to us. No. 9 station was the ideal site and the results obtained here were the standard against which the performance of the other stations could be judged. This station was situated on a small island of deep sandy soil in the bog and yet not of it, for the deep sandy soil made a complete blanket for the peat so that it could not affect matters at all. It was equipped with soil thermometers at the 3-inch and 6-inch depth and with a battery of air thermome-

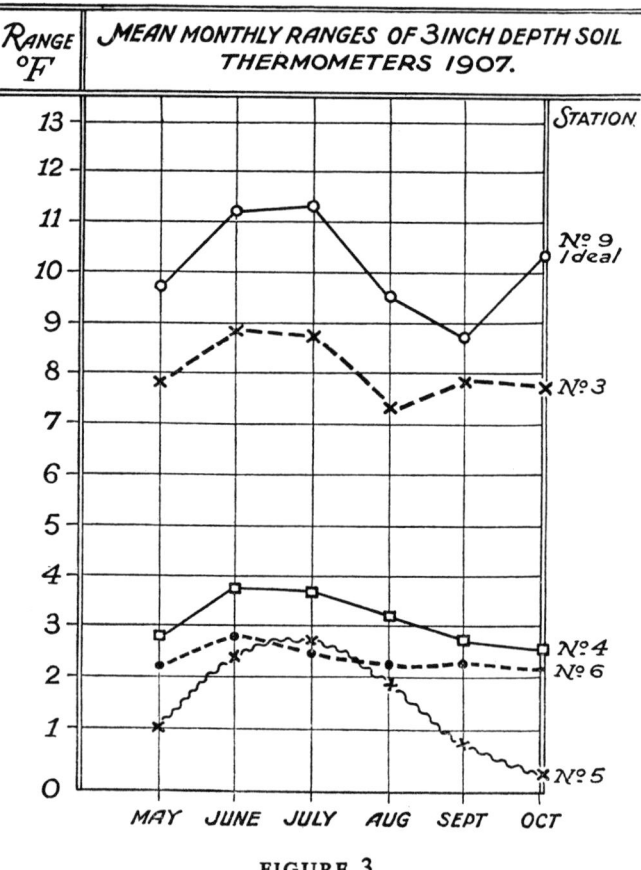

FIGURE 3

ters at the surface, 2½, 5, 7½, 10, 12, 15 and 36 inches above the ground. The other stations were on the marsh, No. 3 on a weeded, newly sanded, and thinly vined plot, No. 4 on a weedy, newly sanded and heavily vined plot, No. 6 on a weedy, sanded nine years ago and heavily vined plot, No. 5 on a weedy, mossy, unsanded, densely vined plot. These were all equipped with 3-inch and 6-inch depth soil thermometers and air thermometers at the surface and 5-inch height above it.

Figure 3 gives the mean monthly ranges of the 3-inch depth thermometers at all these stations. A big range implied a large store of underground heat collected during the day and returned to the surface during the following night to combat frost. No. 3, the weeded, thinly vined and newly sanded plot made a good showing comparable with the performance of the No. 9 plot. But Nos. 4 and 6, the sanded plots, full of weeds and heavily vined, showed up badly in comparison, and No. 5 plot which had never been sanded and was weedy, mossy and densely vined was the worst of the lot. As far as soil temperature and the storage of heat underground were concerned No. 3 plot was undoubtedly the best marsh plot and came not very far behind No. 9 plot, the ideal.

Table 2 gives the mean monthly minima at the surface and in the air 5 inches above for all these stations; these are important as it is between the surface and the 5-inch height that most of the vines are producing their fruit and several degrees of extra warmth in that area is very desirable. Not only did No. 3 plot excel in underground temperature but it also shared the honours with the ideal plot, No. 9, for the highest mean surface and 5-inch height minima, and in July when the berries were ripening and the terminal bud for next year was growing No. 3 plot had the highest surface mean minimum of all the plots. Also the absolute minimum at Plot 3 in July was 34·8° and so the berries were safe all that month from frost while the corresponding minimum at Plot 5 was 27·9° and a damaging frost was likely.

All these results showed very conclusively that over a whole cranberry season the weeding and sanding of the plots did modify very much the drop in the night air temperature, and that it was possible by these means alone to ward off frost in the critical months; as the growers at Mather in Wisconsin must have known these results by the autumn of 1907 it seems extra-

TABLE 2

	Station	May	June	July	Aug.	Sept.	Oct.
SOIL SURFACE MEAN MINIMUM	No. 9 Ideal	40·9	48·7	55·0	52·6	45·6	31·1
	No. 3 Weeded, Sanded	40·4	48·8	55·5	53·7	45·9	28·8
	No. 4 Weedy, Sanded	38·4	45·2	52·3	50·6	44·2	27·6
	No. 6 Weedy, Sanded 9 years ago	38·7	44·0	50·3	49·3	43·1	28·3
	No. 5 Weedy, not Sanded	37·1	43·5	49·8	48·8	43·0	28·0
5-INCH ABOVE SURFACE MEAN MINIMUM	No. 9	40·1	46·8	53·8	51·9	45·3	30·8
	No. 3	39·3	46·3	53·0	51·2	44·6	28·5
	No. 4	38·8	45·7	52·4	50·6	44·1	27·4
	No. 6	38·0	43·2	49·6	48·3	41·7	26·3
	No. 5	37·2	41·8	48·8	47·1	40·7	25·0

ordinary that they did not do more weeding and sanding on the marsh. But many of them objected to the use of sand, because the coarse sand necessary was not available in the immediate neighbourhood, and because they thought natural unsanded peat produced better quality cranberries; and also because the area available for cranberries was almost unlimited, and they contended that they made more money by extending their plots and harvesting poor crops rather than sanding and weeding for better crops.

These arguments were probably fallacious, because by inten-

CRANBERRY MARSHES

sive cultivation it was possible to produce up to 200 barrels of cranberries per acre instead of the average yield of 20 barrels per acre in Wisconsin. But it took the unusually severe frosts of September 1909 to drive the lesson home, when the frosts destroyed 75 per cent of the Wisconsin crop, while the loss at Cranmoor Experimental Station, a few miles away and where weeding and sanding were regularly carried out, was only 2 per cent of the crop. After this disastrous year better cultivation and sanding became much more general in the marsh and the loss of fruit by frost was greatly reduced.

IV

Soils

We have just seen how two inches of sand spread on peat could transform a fruit industry, how much more must a complete change of soil make one garden very different from another. The soil is the home of bacteria which produce a supply of nitrates for the plants from the organic compounds of nitrogen present in a well-manured soil. These bacteria require rather special conditions of warmth, moisture, free aeration of the soil to supply the necessary oxygen, and lime to neutralize the acids as they are produced. These conditions are only found in the upper cultivated layers of the soil, and are more easily obtained in sandy loams than in clays where percolation is too slow, moisture content too high and the supply of oxygen is deficient. Fields of corn turn yellow, especially on the clay lands, when cold and drying east winds cool the soil in the spring, because the bacteria produce too little nitrates in a cold soil to feed the crop; it is only when warmer conditions arrive and the bacteria resume activity that the bright green colour returns to the crop.

Working and cultivating the soil by the thorough aeration they effect, and by the more even distribution of the bacteria in the soil, greatly increase the supply of nitrates, and this cultivation is much more easily done in a sandy loam than in a heavy clay. The hoe is a valuable garden tool, because it kills weeds and helps to produce nitrates; and it covers a lot of ground in a light soil in a very short time. During the summer months a light well-warmed soil, kept well hoed, may produce eight times as much nitrates as the same soil left to itself and never stirred with a hoe.

SOILS

The bacteria like heat and work best at about 100°, they cease work below 41° and above 130°. In 1953 a sandy loam in Edinburgh reached a temperature of 90° or over on twenty days, and of 80° or over on forty-eight days during the summer, and the supply of nitrates produced during these days probably accounted for the local saying that 'our garden will grow anything'. It was only on thirty-three days in the winter, when the temperature never rose above 41°, that the bacteria were completely idle.

Out of the last forty years I have spent twenty in Edinburgh on light loam that would grow anything, and which I could dig very soon after rain and keep my spade clean. These sandy loams were so porous that in winter after heavy rain in the morning, followed by clear sky and sunshine at midday, the temperature at the depth of six inches would rise 5° in an hour as the warmer air was drawn into the soil after the percolating water. In soils such as these the bacteria could never be short of oxygen.

The other twenty years were spent on clay soils, and when I took over my London garden my neighbours all warned me that the clay soil was so intractable that no digging could be done in the autumn or winter without a bucket of water beside me to dip the spade in between every stroke. That sort of cultivation did not appeal to me, as my country upbringing made me realize the hopeless task in the spring of trying to break down the clods that would result from this sort of digging. I did my first digging when the soil was parched in the summer, and with a pick lifted great slabs like paving stones and left them to weather all autumn and winter in the hope of making a tilth for seeds in the spring. After the frosts were over and before the spring rains began I gave the garden a good dressing of short well-rotted compost from the bottom of the compost heap and gradually worked it into the soil.

Before I began operations I had tested the soil to find out what percentage of water made it too sticky to dig, and found it as low as 25 per cent, which of course made it impossible to dig the soil at any time of year except summer. This was all right for the first year when it was not planted with flowers and vegetables, but was obviously out of the question once it was cropped.

So I was delighted to find by the spring after it had been well composted that the soil could contain 35 per cent of water and still be fit to dig. It was of course a heavy clay soil still and had to be treated with respect, but it was now possible to work it in any but the wettest months; and after two years of the same treatment it could contain 50 per cent of water and still be workable. By that time it produced magnificent roses and an herbaceous border that had over a thousand flowering heads on it at one time in June and was a source of envy to my neighbours.

The reason for this enormous improvement seemed to be that the material of the compost heap had the property of imbibing water and expanding in doing so, yet it did not increase the plasticity of the soil but rather tended to open it up and flocculate it. Thus my clay soil was able to hold more water without increasing its plasticity by the addition of the compost.

If only one could combine in one small garden plots of sand, loam and clay so that they all had the same sunshine, rain, humidity and wind, what a lot of really interesting experiments it would be possible to do in comparing these different soils. Thanks to a small research grant I was able to do this on a small scale and had plots made of the natural loam and imported sand and clay alongside each other with the same southern exposure exactly. And in these plots I installed thermometers at the surface and 4-inch depths which were connected by leads to a galvanometer scaled in degrees Fahrenheit in my study. With this delightful apparatus I spent many days and nights taking regular two-hourly readings and accumulated much knowledge and data on the temperature behaviour of garden soils. One of the first results that arose was that my sand plot behaved quite differently to the sanded plots on the cranberry marshes of Wisconsin, because there the peat kept the sand always moist, whereas my plot after a few days sunny weather developed a dry surface mulch.

The capacity of any soil for transforming the radiant energy of the sun into heat at the surface, conducting it down underground by day and returning it to the surface again at night depends on its colour, its specific heat and water content, its conductivity, and the crop that grows upon it. To test the way in

1. Spring flowers in a coppiced wood

2. Frozen spray saves fruit tree from frost damage

SOILS

which colour affected the rate at which a soil absorbed heat, three plots of the natural sandy loam were prepared, one normal, one covered with a dressing of soot to make it black, and one covered with a dressing of lime to whiten it. Thermometers at the 4-inch depth in each plot showed how much heat had been absorbed by each surface.

	Time	Black Surface	Normal Surface	White Surface
May 5th	9 a.m.	55°	50°	48°
	11 a.m.	62°	55°	52°
	1 p.m.	67°	60°	56°
	3 p.m.	72°	65°	60°
	5 p.m.	66°	61°	56°
	7 p.m.	60°	54°	51°
	9 p.m.	56°	50°	47°

The blackened surface obviously got much warmer in the sun as it raised the temperature at the 4-inch depth 7° above the normal and 12° above the white surface at 3 p.m. Also it retained its superiority at night, as the emissive power of radiation of the longer wavelengths radiated by the earth is not affected by colour, and so the black soil did not lose heat more rapidly than the others. The furthest north agricultural research station in the world in Alaska recommends coal dust from the mines to be spread on the frozen soil to thaw it quickly, and so give sufficient growing time for a crop to ripen. The rate at which a material rises in temperature depends on its specific heat. The specific heat of water is 1, the specific heat of every other common substance is less than 1. A pan of milk boils quicker on a gas ring than a pan of water, and many cooks accustomed to boiling a pan of water, by failing to remember the difference, have returned to their kitchen to find the boiled milk all over the floor. Soils contain varying amounts of water when saturated, sand containing least and clay most; so that the specific heat of wet sand is lower than wet clay and the sand will warm up in the sun faster than the clay.

The transfer of heat within the soil depends on the ease with which the heat can be conducted from particle to particle of the soil across the gaps which contain air when the soil is dry and water when wet. Now air has a very low conductivity and completely dry soil with its gaps between the particles full of air may have a conductivity only about a quarter what it can be when it is saturated and the gaps are full of water. There is very little difference between the conductivity of the various soils when they are all equally wet and so finally after taking into account colour, specific heat, and conductivity we arrive at the conclusion that the underground layers in a sandy soil will have a range of temperature one and half times that of loam and twice that of clay so long as all these soils remain equally moist. But as sand dries out rather faster than loam and much faster than clay, after a number of dry days sand will lose its superiority and the underground ranges of all three soils will tend to become equal. These points are well shown in Table 3 for a spell of dry weather in May 1919, and it will be seen that initially the 4-inch depth range in sand was about half as much again as the loam range and twice the clay range, but that after ten dry days the 4-inch depth ranges of all the soils were almost equal.

TABLE 3

Ranges of Temperature at 4-inch depth

1919	*Dry Days*	*Sand*	*Loam*	*Clay*
May 18	0	22·5°	15·0°	11·5°
19	1	24·0°	18·0°	12·0°
22	4	18·0°	15·0°	12·0°
26	8	15·0°	14·0°	13·0°
28	10	14·5°	14·0°	13·5°

Between May 18th and 28th the mean 4-inch depth temperature rose in sand 15°, loam 11° and clay 8°, so clay is a cold soil in summer, partly due to its high specific heat and partly due to the constant evaporation from its surface where it seldom gets dry. From May to September clay is colder than loam or sand at the 4-inch depth by as much as 5°, but is warmer than loam or sand from October to April by about the same amount. So it is generally reckoned a slow starter but a good finisher in the autumn, and this is reflected in the dates of first flowering of several plants and trees on clay and loam soils kindly sent to me by friends in Northamptonshire. They found that aconite, celandine and coltsfoot which flowered in January, February and March were always earlier on clay than on loam, that horse-chestnut flowered at the same date on both soils in April, but that dog-rose and greater bindweed which flowered in May and June were always earlier on loam than on clay.

Within those depths underground between which the roots of our crops are to grow we want in early spring to have a growing temperature of above 42° for several hours in the day even if it falls below that at night, plenty of soil air to encourage the plant-food-producing bacteria, a sufficient supply of moisture not only for current needs, but to tide over any rainless periods without harm to the crops, and at the surface a mulch of some sort to check evaporation since the surface of the soil may be cooled as much as 7° by the evaporation due to a strong wind of low relative humidity. The water needed by most plants for the production of a full crop is so enormous that few fertile soils, unless covered with a mulch at least an inch deep, produce the largest crops they can bear. Thus the conservation of moisture in the garden—especially if the soil is a light loam—is of the utmost importance. With no surface mulch the soil may lose by evaporation up to four inches of rain during the growing period; with this loss the rain needed for a full crop is probably more than the average rainfall in many districts. If so, then some at least of the water required must be raised by capillary action from the soil beneath. The pores of the soil draw up water from below just as a lamp wick draws up oil; but the water must not be too far below the surface just as the oil must not be too far below the burner. As the maximum capillary rise for a light loam soil is probably not more than two feet, the advantage of a layer of turf or compost at about nine inches down to act as a sponge for the percolating rain and return it later to the plant is apparent. That and a surface mulch may save several inches of rain in the first foot of soil ready to give back to the plant in time of drought and so make a full crop possible.

Fortunately the same conditions of high temperature, plenty of water, and oxygen which encourage nitrate production also encourage the growing plant to produce ample roots to absorb the food provided in the soil solution. During its passage through the plant the soil solution gives up its nutrients and is removed by transpiration from the leaves. If we grow similar plants in our sand, loam and clay plots and at intervals pull one or two gently out of each plot, we shall be surprised at the amount of root growing in the loam and clay plots. The total length of the roots of grown plants has been calculated to amount to several miles and each single root may grow an inch or two in twenty-four hours. The conditions in the sand plot are not so favourable and the roots are noticeably smaller; on a hot day in summer the plants on this plot may transpire more water than their roots can absorb from the soil and will wilt.

Then the natural reaction of the thoughtless gardener is to rush for the watering can, fill it at the main tap, and hastily water the wilting plants, and then be surprised and disappointed that the plants do not recover but only wilt all the more. The reason really is obvious, for on a hot day the soil may be at a temperature of 80° to 90°, while the tap water is at 55° to 60°. Since absorption of water by the plant slows down as the temperature of the soil falls, we have now made it impossible for the roots to supply the plant with as much water as they did before. The wise gardener avoids this trouble by always having a can full of water in the sun all day ready for use, or by taking his water from a water butt kept supplied by the roof of his garden shed.

There are dozens of ways of ill treating a soil—especially a clay—and the old hand, who in his early days no doubt committed every sin, now avoids them all. Notice how your neighbour, who has a clay soil, walks on a board rather than on the soil, when he sows his peas or sets out his cabbages and how carefully he restores the tilth with a gentle raking where the board has flattened it. Perhaps you say to yourself, 'fussy old beggar', and pass on. But go home and plant a bit of your own clay soil your own way, and trample all over it in doing so, and leave it untouched when you have finished. The first time it rains your footmarks will be puddles of water which the sun and rain will evaporate in no time, and your plot will probably suffer from drought. Years ago I purposely adopted this wrong method on my clay soil and lost three-quarters of my young cabbages from lack of water in the drought. In a properly planted plot alongside only two cabbages died.

Not only does wrong handling of a clay soil affect its immediate future, but a really bad case of mismanagement in wet weather may take a year or two to put right. This, on a farming scale, may be quite a disaster and the subject of much comment in the village public house. Years ago our wagoner heard some news there and passed it on to his mother who came to work in our home every day. So Father knew all about it by midday, when the young farmer whose orders had caused the calamity came to ask his advice. He had ordered a field to be ploughed when it was not fit, so that it would never be fit for sowing that autumn, and he had left his bullocks too long on a clay meadow in wet weather. This might not have been disastrous, but a stray dog had got into the meadow, and the bullocks, in their efforts to chase it and emulate its quick turns, had galloped and slipped and slithered all over the meadow so that it would take a couple of years to heal its wounds. He asked Father what he was to do, because now his men thought he was a fool. 'I know most of your men', Father replied, 'and they are good countrymen. If you had asked your wagoner about ploughing the field he would have advised you against it; if you had sent your cattleman to see your meadow he would have suggested taking your bullocks out of it. Until you have learnt your job you should take your men into your confidence.'

'But I can't do that now, they would tell me wrong just to make a fool of me.'

'Wrong again,' said Father, 'a good countryman would think it a mortal sin to maltreat a field and would never advise doing it, not even to score off his greatest enemy.'

That sentence of Father's which I heard as a boy I have remembered all my life, and when in doubt on country matters I have always asked a good countryman.

V

Taking Earth Temperatures

As children we looked forward to the spring months in the country when milk and cream were plentiful, because we were allowed to make cream cheeses, which had to be ripened in an even cool temperature. So we wrapped them in cheese cloth and buried them about a foot under the grass in the garden and left them there for several weeks. We knew that the temperature at that depth under grass was a good deal cooler and did not change so much as under the bare soil and so was ideal for ripening the cheeses.

And when we found that the rabbits were using their top burrows again, that our hedgehog had come out of his hibernation, that the wormcasts ceased to show on the lawn, and the mole-heaps got bigger as the mole went down deeper after the retreating worms, then we knew that the tide of earth warmth was flowing again and it was time to get our packets of seeds for our gardens. To a country child interested in every bird and beast it all seemed so plain; the walls of our home were made of bricks dug out of one of the fields still called the Brickyard, our butter and meat and milk were all kept in a dairy dug several feet into the earth, and so was cool in summer and warm in winter like the rabbits' deep burrow. Earth temperatures were so much a part of our everyday lives that we should have been surprised at anyone who knew nothing about them.

As a schoolboy I put this knowledge to good purpose when I did some work for a neighbour who wanted to find how to stop spring frosts from damaging his fruit blossom. His orchard was on the slope of a hill and at my suggestion he ploughed up the grass beneath the trees in one half and left the other half under

TAKING EARTH TEMPERATURES

grass as before. I found, as I expected, that the temperature of the ploughed soil was so much warmer than the grass both by day and night that the air temperature above it was appreciably higher than over the grass even in the early hours of the morning when frost was most likely. So the orchard was kept ploughed and the severity of the frosts was much abated. In later years, when I became interested in micro-climates, it was the memory of this first successful piece of work that encouraged me to spend some time taking earth temperatures and so putting actual figures to the rather vague ideas of temperatures underground that I had taken for granted in my youth.

The warmest and coldest times of year are not normally at midsummer and midwinter when the sun is highest and lowest in the sky, but at August and February. This is due to the fact that air is hardly heated at all by the sun, but takes its temperature mainly from the earth and the sea, and the maximum earth temperature is in August and the minimum in February as we have seen in Table 1. So air temperatures near the ground are largely controlled by the temperature of the soil surface.

That part of the sun's energy which is not absorbed by the air or reflected from the earth's surface is absorbed by a thin layer of about two feet of soil and the heat thus provided is conducted downwards and stored so long as the heat at the surface is greater than the heat in the deeper layers of the soil. It takes some time for the wave of conducted heat to travel downwards and at the 4-inch depth the time lag of maximum temperature by day is often as much as $3\frac{1}{2}$ to 5 hours, according to whether the soil is wet or dry, and so the maximum often arrives at the 4-inch depth at from 5 to 6.30 p.m. At night, when it is calm and the sky is clear, the surface cools quickly by radiation and the stored heat underground is conducted back to the surface with the same time lag, so that the minimum at the 4-inch depth may arrive about 7 to 8 a.m. in the summer and 11 a.m. to noon in the winter. I used to find the lag of maximum and minimum temperatures underground very convenient, as I could set the index of the maximum thermometer as I went to work and of the minimum as I came home in the evening, and so

TAKING EARTH TEMPERATURES

ensure that I did not miss a day's record or duplicate yesterday's readings if I forgot to reset the index.

A very useful type of soil thermometer is the plain glass tube with the scale engraved on the glass and the bulb on a 4-inch long stem bent at right angles to the scale. This ensures that, when the thermometer is put into the soil with the scale just resting on the surface, the bulb is exactly four inches below the surface. A pair of these with mercury for the maximum and spirit for the minimum will be needed. For the surface temperatures a pair of plain straight maximum and minimum thermometers with the scale engraved on the glass will be needed and the correct exposure of these is important. It is no good just laying these thermometers on the top of the soil as they will pick up all the changes that take place from minute to minute as the sun goes in and out of a cloudy day. These transient changes are not transmitted to the 4-inch depth and to avoid picking them up it is best to press the bulbs of the thermometers into the surface and just cover them with a thin layer of soil. Suitable scales for the surface thermometers so as to be able to cope with an extremely hot day in summer and a very hard frost in winter might be from 0° to 120°, while for the 4-inch depth the same range can be used and then the four thermometers will be interchangeable for a short time in an emergency.

Once installed the thermometers can be left for months and will stand up to hard frosts without any damage. Never use cheap thermometers for micro-climate work and especially avoid those mounted on a wooden frame with the scale on the frame; they suffer from so many disadvantages that they are quite useless in the open in a garden; the frame warps, the scale fades out and becomes illegible, parallax makes them difficult to read accurately, and the stem and bulb is liable to slip through the holding clips and move up and down the scale. A good instrument may cost 30s. to 40s., but with careful handling will last for years and give really accurate results. As a general rule accuracy to the nearest degree is all that is necessary, but it is important to ensure that the reading is correct to that degree of accuracy.

In comparison with the erratic variations of air temperature

3. A frost hollow

4. Fruit ripens well on a south wall under a wide coping

TAKING EARTH TEMPERATURES

the regular pulsations of temperature underground follow well-known laws for amplitude and time lag according to depth, and the chief factor in this orderly process is the conductivity of the soil, which largely depends on the amount of moisture it contains. Anyone who has had to measure soil conductivity in a laboratory knows that it is a tedious and troublesome experiment, and this probably explains the paucity of published results for various soils with different degrees of wetness. And when we consider that every summer we are dealing with a soil whose moisture content varies from that of a dry mulch at the surface to a very moist soil below, we might well despair of ever being able to compare the varying conductivities from hour to hour and day to day. And yet this is exactly what the soil temperatures do for us.

Now that I had the means at my disposal I decided to try to discover the whole range of temperatures in a soil at the surface and 4-inch depth between just after heavy rain and a state of as complete dryness as our rather infrequent heat waves would allow. For this, by good luck, I had only a few weeks to wait before I got exactly what I wanted. The complete results are given in Table 4, which is a record of the behaviour of a sandy loam at the surface and underground during a spell of sixteen consecutive dry days, of which fourteen were sunny, after a heavy downpour of rain.

Several points of interest arise from the figures in this table. The surface range rose more or less steadily from 36° to 52°; this was not because the sun was getting so much hotter, but because a dry mulch was forming on the surface of the soil and the enormous loss of heat due to evaporation of moisture was being reduced. But in spite of the rise in the surface range, the range at the 4-inch depth remained more or less steady at about 15°, because the conductivity of the dry soil in the mulch was so much reduced that the extra heat in the surface was not conducted downwards. In fact, a surface mulch of two inches acts as a most efficient thermostatic control of underground temperatures.

Although in America deep mulches are constantly advocated,

and F. H. King in his *Physics of Agriculture* has proved the value of a 4-inch mulch, it is doubtful if mulches of over two inches are of much value in this country where long periods of dry

TABLE 4

The Thermostatic Control of a Surface Mulch

Date	Surface			4-inch Depth			Depth of Mulch	Weather
	min.	max	Range	min	max	Range		
May 18	45	65	20	51	60	9	0	Sunny after Rain
19	42	78	36	50	65	15		Sunny
20	42	78	36	50	65	15		,,
21	42	62	20	50	58	8		Overcast
22	47	85	38	52	67	15	½ inch	Sunny
23	50	90	40	54	69	15		,,
24	50	94	44	56	72	16		,,
25	50	98	48	56	72	16	1 inch	,,
26	43	93	50	52	68	16		,,
27	45	95	50	54	69	15		,,
28	44	94	50	54	69	15		,,
29	50	75	25	56	63	7		Overcast
30	44	95	51	55	70	15	2 inch	Sunny
31	45	95	50	56	70	14		,,
June 1	48	100	52	57	72	15		,,
2	50	102	52	58	73	15	2½ inch	,,

weather of over sixteen days' duration are comparatively rare. A 2½-inch dry mulch appears to reduce the value of the conductivity of the soil down to the 4-inch depth to about half its value when it was uniformly moist.

While the surface temperature ranged from 42° to 102° and the bacteria on the surface for part of the time were more or less inactive by being too cold at 42° and too hot at 102°, the temperature at the 4-inch depth lay between 50° and 72° and for the

TAKING EARTH TEMPERATURES

whole of the period the bacteria were producing nitrates steadily at a more or less even rate thanks to the thermostatic action of the mulch. Now that we have seen how the underground layers of the soil gain their heat in the summer we must consider how they lose it in the winter. Just as the sun gives radiation on many wavelengths to the earth during the day, some of which are largely absorbed by the water vapour in the atmosphere giving us days when the sky has a whitish tinge instead of the blue of a day of dry air, so the earth also reradiates at night on many wavelengths some of which are absorbed and radiated back by the water vapour. It was Anders Angström, the Swedish physicist, who worked out the relation between the final outgoing balance of radiation from the earth at night and the relative humidity when there was a cloudless sky and no wind.

In Table 5 I have attempted to make a balance sheet in British Thermal Units—the amount of heat needed to raise one pound of water 1° F.—of the soil heat on several nights in winter. For the credit side I have used the upward conduction to the surface from the 4-inch depth, and the Latent Heat of Dew or Hoar Frost when these were formed. For the debit side I have used Angström's figures of the outgoing radiation from the value of the Relative Humidity. On striking a balance, if the account was overdrawn a frost was inevitable.

In the course of these experiments I learnt what I believe is one of the great truths of work on micro-climates—that mean values are of very little use. Data about climate are mostly published in the form of monthly or daily means, and two March months, when I had animals and birds under observation, had exactly similar mean temperatures. The first had a very small daily range of temperature and a mean equal to the average, the second had fifteen frosty days at the beginning and sixteen warm days at the end and a mean equal to the average. But the animals and birds I was observing behaved very differently in these two months. Also, when years ago Henry Mellish of Hodsock Priory sent me details of dates of first flowering of coltsfoot at Hodsock, we could find no relation between the dates and the mean temperatures; but a very exact connection between the date of flowering and the number of frosts when snow was not lying in the two months previous to the appearance of the flower. In fact, ani-

TABLE 5

Balance Sheet of Soil Heat in British Thermal Units per Square Foot

Date	Creditor			Debtor	Balance	Remarks
	Upward Conduction	Latent Heat	Total	Radiation		
Dec. 4	288	148	436	420	+16	Ground warm after mild spell. No frost
15	412	148	560	540	+20	,,
Jan. 19	108	360	468	500	−32	Ground cold. Frozen to 1 in. depth.
Feb. 9	110	360	470	520	−50	,,

mals, birds and plants appear to react to the extremes and not to the means of climatic data, and in any study of micro-climates we must be on our guard against assuming that mean values are of any use.

In this country the depth of winter is about the middle of February, and by that date the underground layers of soil in which our crops are to grow have been reduced to about freezing point and at a depth of 4 inches the range of normal temperature is from about 30° to 34°. If we consider the minimum temperature for germination and growth of most crops, and for the activity of those bacteria in the soil which supply these crops with food, as about 42°, the 4-inch depth has to rise about 10° in temperature before growth will start. This would be a fairly rapid process under the influence of stronger sunshine, but for the frosts, snow and cold rains which are characteristic of our early spring. In consequence it is the end of March or the beginning of April before the soil arrives at a growing temperature in most years. After that the rapidity with which growth proceeds depends on the accumulated hours of temperature that the soil can show above the base level of 42°, and the number of hours

TAKING EARTH TEMPERATURES

at a time that this temperature persists. If the temperature at the 4-inch depth remains above 42° for at least a few hours daily growth will take place even though the soil falls considerably below that at night.

In March and April 1949 there were two spells of very different weather which by their influence on the 4-inch depth temperature showed that it was the overcast weather with bright intervals and overcast nights that most rapidly increased and maintained the underground temperature, and not, as might be expected, the bright sunny weather with clear sky and low relative humidity. Daily temperature changes are in the main periodic, the early morning minimum and afternoon maximum occurring very regularly in spite of weather changes. But there are occasions, associated with a strong westerly air stream over these islands for several days, when the interval between minimum and maximum may be as much as 100 hours. Figure 4 shows that it is this extreme weather rather than the ordinary short period changes that brings spring to life underground.

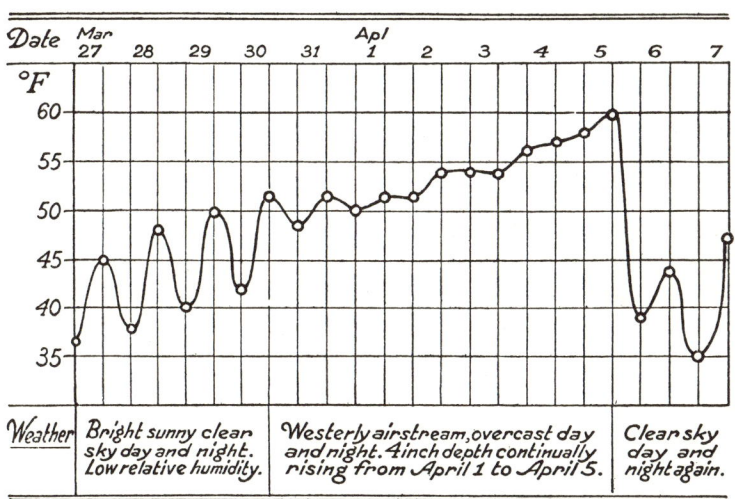

FIGURE 4

The effect of Weather Changes on 4 inch Depth Temperatures

VI

What Happens to the Rain?

Change is the spice of life. Yesterday, after several days' fine sunny weather, the surface thermometer was in the 90's, the two-inch dry mulch looked grey and tired, the evaporimeter was working furiously, the plants were wilting, and I could imagine the bacteria down below saying that it was a long time between drinks. To-day it is raining, the thermometer does not move, the evaporimeter is asleep, the mulch has gone and the soil is brown and shining, and I am sure the bacteria are working all the better for their drink. I can almost hear them chuckling because the rain has gone to the old man's head; perhaps they are right, for there is something extremely satisfying in watching much-needed rain refresh a garden.

What will happen to the rain that the garden is enjoying to-day? There are four ways it can be disposed of, namely by surface run off, by evaporation from the crop and the surface of the soil, by transpiration by the growing crop, and by percolation to lower levels and eventually into the drains or water-table. In a well-kept garden where the hoe is used frequently surface run off should not exist; but on our farms it may be considerable in these days when most of the harvest operations are done by tractors, and the farmer likes to plough up his hard-pressed stubble as soon as possible after harvest, and thereby conserve in the top few feet of soil as much moisture as may be equivalent to two inches of rain.

Evaporation is the greatest thief of moisture so let us think for a moment about some means of keeping a record of these thefts. The evaporation of water from a water surface has been measured for many years at Camden Square in London and the 35-

WHAT HAPPENS TO THE RAIN?

year average amounts to 13·5 inches out of a rainfall of 25 inches. This value is of little real interest as a free water surface is very different to a soil surface and does not grow a crop. The drain gauges at Rothamsted were made by building cement walls round a block of earth and then without disturbing the block gradually introducing a perforated iron plate to carry the soil. Here the 35-year average shows that the soil blocks which grow no crop evaporate about 14 inches out of a rainfall of 29 inches. If we consider the six months April to September the comparable figures are 10 inches evaporated out of 14 inches rainfall. This again does not approach reality owing to the absence of a growing crop which, quite apart from the moisture it uses in transpiration, shades the surface from the sun and shelters it from the wind and so reduces considerably the evaporation from the soil surface.

A friend of mine determined to get close to reality by making a drain gauge with an imitation grass crop on it; he hoped thereby to find out roughly how much evaporation went on in a grass turf. In his orchard he had an old stone drinking trough dug into the ground and arranged for the outlet pipe to be led to a sump where percolation could be measured daily. He filled the trough with soil and after a period to let it settle he put dead turves all over the surface and then covered it with his imitation grass mat made of straw.

Whether this simulated grass crop had the same effect on evaporation as a real crop it is impossible to say; but certainly it reduced the evaporation considerably, and during the six months April to September, out of a rain fall of $14\frac{1}{2}$ inches, 6 inches were evaporated and $8\frac{1}{2}$ inches percolated. The following year, during the same six months, with a rainfall of $12\frac{1}{4}$ inches just under 5 inches were evaporated and $7\frac{1}{4}$ percolated. So that it seemed as if a reasonable value for the evaporation from grass turf was about two-fifths of the rainfall.

At the same time I made a number of observations of the amount of rain that could fall and be caught and evaporated by the covering crop without wetting the surface of the soil at all. This of course varied with the kind of crop and the size and shape of the leaves, but as a general average it was safe to reckon that falls of up to two-hundredths of an inch were evaporated as they fell and never reached the soil. This corresponds to what the farm workers in Northamptonshire used to call the 'taith' drop, which may have come from the Welsh 'llaith' meaning damp. For the taith drop they would never even put on their jacket, which may account for the prevalence of rheumatism in the villages.

For general use in my garden I find a simple form of evaporimeter very useful. Nothing elaborate is necessary as we have no need to be really accurate, all we want is some means of knowing the different rates of evaporation in different types of weather. My instrument is a plain straight-sided jam-pot with a loop of wire, which when draped with a piece of well-washed flannel as a wick, fits snugly into the mouth of the jam-pot and prevents the loss of any water except through the flannel. The scale on a piece of paper stuck on the outside of the pot has to be graduated empirically allowing for the space taken up by the wet flannel wick. Fitted into a wooden box with a hole in the lid through which the top of the jam-pot and wick project slightly, this simple instrument when exposed in the open gives a good idea of the evaporation over several days of fine, or cloudy windy weather. The flannel wick needs washing at intervals to remove the grease, soot and dirt that it picks up in the garden, otherwise it is completely trouble-free. On breezy sunny days in the summer, readings of about a fifteenth of an inch a day are common, and on exceptional days of low humidity a tenth of an inch may be recorded; these are the good drying days for the household washing because for quick drying it is essential that the dry air that is sucking the moisture out of the clothes should be replaced by fresh air as soon as it has done its job. A windy day without sun is a better drying day than a sunny day without wind.

The difference between the air temperature over open soil and the temperature of the soil surface, kept damp by watering if necessary, is another useful measure of evaporation. It takes into account all the variables of sunshine, humidity and strength

WHAT HAPPENS TO THE RAIN?

of wind, and a difference of about 6° corresponds to about a tenth of an inch a day on the evaporimeter and a time of about thirty minutes to dry a washed and well-wrung face towel on the line in the garden. When the difference was 3·5° the towel took nearly an hour to dry as the evaporation was obviously only about half as effective.

The chemical changes going on in a plant in bright sunlight might produce a rise of temperature of over 20° in a minute, and in a few minutes death would ensue, if the temperature were not kept in control by the evaporation of moisture from the leaves in transpiration. Transpiration also promotes the ascent of sap in a plant and the influx of water and mineral salts through the root hairs in the soil. It is increased by sunlight, warmth, wind, and low relative humidity, so that it is greatest in the summer months when most of these conditions prevail and when the plants are growing luxuriantly.

It is a pity that the amount of moisture transpired by various crops is not known exactly, and that the transpiration ratios for them—or the number of tons of water per acre, to produce a ton of dry matter in the crop—are given very different values by different observers. The values of the ratio for peas, for instance, ranges between 235 and 788, and for grass from 300 to 1,076. Nor is it helpful in expensive books on plant physiology to find that the causes that produce these large variations such as variations of light, heat, wind, humidity, water supply, and mineral deficiencies in the soil are not discussed but left as an exercise to the student. If we could present to the farmer and gardener a simple estimate of how all these factors affect the transpiration ratio, we should make it easier for them to decide how to alter their crops to give a maximum yield with the soil they have, the climate they normally enjoy, and the manuring they can afford.

With a view to seeing how two amateurs' figures compared with those of the expert, I got my friend to grow a crop of real grass in his stone water trough, while I prepared an experimental plot for potatoes. He sowed timothy seed in the autumn, cut it at the end of April when there was no percolation and then

WHAT HAPPENS TO THE RAIN?

left it to grow into a hay crop until the end of July. In those three months he had 8¾ inches of rain and no percolation at all as the rain was well distributed and no thunderstorms visited him. On the assumption that two-fifths of the rainfall was evaporated, as in his previous experiments, this left 5¼ inches of rain transpired to produce what turned out to be a hay crop of about 1½ tons to the acre of timothy hay of which 84 per cent was dry matter, so that his crop had transpired 5¼ inches of rain in producing 1¼ tons of dry matter. As 5¼ inches of rain represents about 530 tons to the acre the transpiration ratio of his timothy grass was about 424, and lay rather near the lower limit of the values given by the experts.

The potato experimental plot was 30 feet long and 15 feet broad and sloped down to a small stream that was usually dry in the summer; drains of open metal pipes covered with stones were connected to a central pipe at a depth of 2 feet which ran down the centre of the plot and discharged into the stream. The plot was dug early in April and manured with a good layer of manure between the 4-and 6-inch depths; well-sprouted potatoes were planted on the top of the manure with artificial potato manure at distances of about 18 inches from set to set and row to row, so as to give a uniform distribution of vegetation cover over the whole plot. The soil was a sandy loam well manured and free working and quite ideal for a good crop.

For a week before planting the potatoes the drain had run freely, but had ceased running by the date of planting; the actual cultivation and manuring of the soil, although causing some loss of moisture by evaporation during the operations, no doubt rendered the soil more capable of holding water owing to the layer of manure it contained. It was expected that a considerable amount of rain could fall without causing the drain to run again, and in fact 1½ inches of rain fell between the date of planting and May 18th without causing the drain to run. On May 18th ⅛ inch of rain fell and this caused the drain to run very slightly and the soil on that date was saturated down to the depth of 2 feet; at this date the potatoes were well above ground and transpiration was assumed to have begun. *The drain did not*

WHAT HAPPENS TO THE RAIN?

run again until October 30th, and the estimated figures showing what had happened to the rain from May 18th to October 7th, when the crop was dug, were as follows:

Rainfall..	9·5 inches
Interception and evaporation	3·8 inches
Drainage	Nil
Balance for transpiration ..	5·7 inches

The crop when dug worked out at 8 tons to the acre and as a quarter of this was dry matter the potato crop had transpired 570 tons to the acre of rain in producing 2 tons of dry matter. So their transpiration ratio was 285, compared with the figures of 250 to 636 of the experts. As soon as the crop was dug and transpiration ceased fairly heavy rain on October 19th caused the drain to run for the first time for twenty-three weeks. The average rainfall for the period was about 11·0 inches, and as other conditions were all favourable I think the rainfall was the limiting condition, and had it been up to the average the crop might have worked out at 10 tons to the acre.

All water in the soil is not subject to gravity and, fortunately for the farmer and gardener, their crops do not have to lead a hand-to-mouth existence fighting for water before it disappears into the lower depths out of their reach. When a soil is dried in air at ordinary temperatures a certain amount of moisture still remains in it, which depends on the humidity of the air and the type of soil. We can test this by weighing a sample of dry soil in the laboratory when we shall find that it weighs more on damp days than when the weather is dry and sunny. This minimum of water that can seldom be got rid of is called hygroscopic moisture; its amount is found to depend on the amount of colloidal matter that the soil contains, so that the hygroscopic moisture in a sandy loam is only about 2 per cent, in a clay loam 6 per cent, while in a heavy clay there may be over 10 per cent. It is not available to plants and growth ceases and plants wilt when the water content of the soil is about $1\frac{1}{2}$ times the hygroscopic moisture.

But when more water is added the soil colloids—which include any manure or compost incorporated in the soil—absorb

WHAT HAPPENS TO THE RAIN?

this water and swell considerably in the process. Though this moisture is held very tenaciously against mechanical forces such as pressure or gravity, it is available to plants and is their chief source of moisture. After this colloidal moisture reaches its maximum any further addition of water causes percolation to begin.

So the balance of the rainfall after we have accounted for run-off, evaporation, transpiration, hygroscopic and colloidal moisture, should normally percolate under the action of gravity through the pore spaces of the soil, and also through cracks, worm tracks, and the passages where roots have decayed. This normal percolation is an orderly movement of water down to the drain level, where some is lost to ditches, streams and rivers, and after the drain level is passed, to the water-table which is the level at which water stands in wells. In winter months as much as four-fifths of the rainfall percolates to the water-table, but as we have seen during the summer months the percolation through soil growing a crop may be negligible, and even in soil which grows no crop only about one-fifth of the rainfall reaches the water-table.

When the ground has been dried to some depth in the summer and is covered with a surface mulch, percolation is sometimes hindered by the air within the soil being unable to escape when the top layer has been thoroughly wetted. We then have a wet surface zone, a dry zone full of air and having no continuous film of wet soil grains to lead the water down to the wet zone beneath. It takes some time for small displacements of the air to be set up and so provide a path whereby the wet zones in the surface and subsoil can make connection and percolation can begin. Sometimes this connection is never made and the surface water is eventually wholly evaporated, so that the first summer rains that end a drought are often of little benefit to the crops. Gardeners are often laughed at by the ignorant for watering their gardens as soon as there is news of the break up of a drought, but as we have seen there is method in their madness.

In coarse-grained sandy soil percolation is very rapid and the soil dries quickly after rain; in a clay loam, which contains a lot

WHAT HAPPENS TO THE RAIN?

of fine grain and colloidal material, percolation is slowed down and the soil dries slowly; in a heavy clay soil percolation may be so slow that the upper layers remain saturated all winter and the drains only run very slowly in spite of the mass of water the soil contains. The water that has percolated down to the water-table can—under certain conditions—rise again in time of drought, just as oil rises in a lamp wick if the burner is not too high above the level of the oil. Dr. B. A. Keen, at Rothamsted, found that in heavy loam a capillary rise of 32 inches was possible; this was reduced in sandy loam to 28 inches, and in a coarse sand to 14 inches. So that unless the water-table was less than these distances from the roots of the plants in the various soils this capillary moisture would not be available to them.

The water that runs away in the drains into streams and rivers and that which remains in the water-table and wells contains a considerable amount of dissolved manure and bacteria, and probably requires treatment before it is fit for drinking. But for irrigation purposes it is admirable, and has the double advantage of higher temperature than the land surface in the early spring and good manurial value. At home we always flooded the meadows in turn every spring, and father considered it was folly not to use the manurial residues from his neighbour's lands upstream which flowed so conveniently through our meadows. And certainly the enormous crop of hay that resulted from the early flooding bore witness to the good manurial value of the river water.

VII

Nature's Way

An oak and hazel wood which has not been coppiced for at least ten years is an ideal site to study nature's way, which has been epitomized in the saying, 'Warmth, water, and air first'. Not the fierce heat of the sandy soil in the midday summer sun, or the bitter cold of its surface in the frosts of winter, but a more genial climate that lies somewhere between these extremes. And while the heavy rain rushes down every furrow of the neighbouring grass field on its way to the brook at the bottom, in the wood it seeps quietly into the soil through the cover of grass, moss, and leaf mould which acts as a sponge to catch the rain and waste not a drop. Somewhere below the moss lie the creeping root stocks of the wood anemone and dog's mercury, the tubers of the lesser celandine and cuckoo pint, the bulbs of the bluebell, and the roots of the primrose. All these provide channels to lead the rain down to the subsoil and eventually to the water-table.

My wood was too far from home to allow me to take daily maximum and minimum temperatures, but I was fortunate enough to be able to do this for a few days at mid-summer and mid-winter, and so could compare the warmth of the soil under the grass and moss and fallen leaves with the open garden soil on the hottest and coldest days of the year. When the surface was 107° in the garden it was 68° in the wood, and on the morning when the surface in the garden fell to 20° and the soil was frozen to a depth of four inches, the soil was unfrozen under the moss, grass and leaf mould and the surface was 33°. So the few inches of forest cover made a range of 35° for the year against 87° in the garden. To test the porosity of this forest cover I collected a box of it and compressed it to about two-thirds of its natural volume, and then found that even in this compressed form the rate of percolation of water through it was five times the rate of a similar box of uncompressed garden soil. No wonder the wood in the spring was a marvellous carpet of bluebells, anemones, dog's mercury and celandine, for nature had exerted all her wiles to give their roots a medium which from the point of view of warmth, water and air was ideal for them.

These results were so interesting that I decided to use the wood soil for a much more extensive series of experiments, and I had a load of soil from the wood complete with its covering of grass and moss brought down to the garden and made a sizable bed with it in which I could measure at my leisure its thermal efficiency and compare it with the efficiency of many other poor conductors when laid on the surface of the soil. As long ago as 1883 Carl Eser found that between July 12th and August 12th bare soil lost nearly 6 grams of water per square centimetre, or roughly about 5 inches of rain, by evaporation; a 2-inch layer of chopped straw and a similar layer of beech leaves both reduced the evaporation to half an inch of rain. H. Burger in 1926 made some studies of the effect of forest cover on the water and air capacity of the soil, and found that water was taken up by forest soil about twenty-five times as quickly as by a permanent meadow, and that the air capacity of the forest soil was seven times that of the meadow soil.

But so far as I could find, very little work had been done on temperatures under various poor conductors. As the depth of the moss and grass from the wood was about two inches I decided to have a row of plots with the same exposure and cover them with two-inch layers of loose soil, ashes, farmyard manure, leaves, short turf, and moss, grass and leaves from the wood and compare the temperatures under these layers with the temperature two inches down in open firmed soil. The experiment lasted for four months and some of the results are given in Table 6; in this four months I was fortunate enough to have two quite long spells of frost during which the temperature of the surface of the open soil fell to 20° and the soil itself was frozen to

NATURE'S WAY

the 4-inch depth, so that all the poor conductors were pretty severely tested. Very early in the experiment it was apparent that the thermal efficiency of the best of these overcoats was due to the large amount of air they contained and the quantity of dew or hoar frost that was condensed in them every night, and the leaves, short turf, and moss and grass were particularly noticeable in this respect. So as a check I arranged a plot with leaves set vertically and not horizontally as they always lay in the woods; this of course reduced the amount of trapped air to a

TABLE 6

Minimum Temperatures under 2 inches of various poor conductors

Date	Open Firmed Soil	Loose Raked Soil	Ashes	Farm-yard Manure	Leaves	Short Turf	Moss Grass Leaves	Remarks
1918	deg.	deg.	deg.	deg.	deg.	deg.	deg.	
Nov. 8	38·0	39·5	41·0	41·5	44·5	46·0	48·0	Open soil frozen to 1 in. Nov. 15, to 4 in. Dec. 26, to 4 in. Feb. 17. Soil under short turf and grass moss and leaves *never froze* though surface of open soil fell to 20° and grass minimum to 15°. Soil under leaves frozen to ½ in. on Feb. 10.
9	34·5	36·0	38·0	40·0	42·5	43·0	45·0	
11	36·5	37·5	40·0	41·5	43·5	44·0	46·0	
12	30·0	32·0	35·0	37·0	38·0	39·0	40·0	
13	30·0	32·0	35·5	37·0	38·5	39·0	40·0	
15	31·0	33·0	35·0	37·0	38·0	39·0	40·0	
Dec. 15	30·5	32·5	34·5	36·5	38·0	39·0	41·0	
16	30·0	32·5	34·0	36·5	38·0	39·5	41·0	
17	30·0	32·0	34·0	36·5	38·0	39·0	41·0	
18	30·5	32·5	34·5	36·5	38·0	39·0	41·0	
19	29·0	31·5	32·5	34·5	36·5	38·0	40·0	
20	26·5	29·5	30·0	32·5	33·0	34·0	36·0	
21	31·5	33·5	35·5	38·0	38·0	40·0	41·5	
22	29·0	32·0	33·0	36·0	36·0	38·0	40·0	
26	29·0	32·0	33·5	36·0	36·0	38·0	40·0	
1919								
Feb. 8	29·5	32·0	34·0	36·5	37·0	39·0	40·5	
9	24·5	28·0	29·0	32·0	32·0	33·5	35·0	
10	22·0	26·0	26·5	29·0	30·0	32·5	34·0	
11	26·0	29·0	29·5	33·0	34·0	35·5	36·5	
17	28·0	31·5	32·0	35·0	35·5	36·5	38·5	
Average increase of temperature over open firmed soil.		2·4	4·0	6·2	7·5	8·8	10·5	

minimum, and the plot with vertical leaves was not nearly so

NATURE'S WAY

thermally efficient and fell back into the same class as the ashes.

When I was a boy I made some pocket-money by catching moles for the neighbouring farmers and learnt my trade from an old professional mole catcher who taught me that anyone can catch moles in the summer, but only the masters of the craft could catch them regularly in the winter. Because when the ground was frozen the mole could not make new runs under the frozen soil as he was unable to break through the frozen surface to push up the heaps of earth he had dug. So he lived on the worms he had bitten and stored near his home or he made new runs under the long grass by the hedgerows or under trees where only the hardest frost froze the ground. Here, or in his water run which he was bound to use every day, one had to set the traps if one was fortunate enough to find the water run and clever enough to set a trap in hard frozen soil. So as early as 1892 I knew about the efficiency of grass and moss although it was not until twenty-six years later that I proved it with thermometers and showed that on 10th February 1918 the temperature under natural moss and grass was 12° higher than under open soil in a hard frost.

On the banks of ditches, under the lee of hedgerows, and in the woods protected from the wind and radiation by the canopy of the trees and a covering of moss, grass and leaf mould, the roots of spring flowers lie untouched by frost all through most winters. A spell of warm wind and rain any time during the winter will make them put up their leaves and we find them sometimes in flower when winter has returned with all its vigour. Deep-rooted plants which grow in these sheltered places have an enormous advantage; at the depth of their roots the winter cold never penetrates and probably the temperature there never falls below 42°, so that any slight increase in warmth will make them start growth. The roots of coltsfoot go down to a depth of six feet and its flower, which comes before its leaves, can only be delayed by being unable to force itself through the frozen ground. After many years study of the date of first flowering of coltsfoot in places where frost can penetrate into the soil, I have found a very distinct connection between the date of flowering and the number of days when the ground was frozen without a covering of snow after January 1st. One particular patch of coltsfoot in the edge of the wood, where frost hardly ever penetrated, flowered with great regularity year after year about the beginning of February, while another patch on bare stony ground often frozen to a depth of four inches, was generally a month later in a hard winter.

After woods the banks of ditches sheltered by good sized hedgerows are probably the next best site for early spring flowers. I have taken temperature readings of these sites and have found on calm clear nights in winter a difference of as much as 7° in favour of the bank and ditch over open ground, and as much as 4° on wet and windy days in favour of the lee side of the hedge over the windward side due to shelter from the wind and less evaporation. And it seemed as if the difference on calm frosty nights was due partly to the hedge cutting off half the open sky to which all things radiate away their heat at night.

So amongst permanent winter overcoats for the earth, probably thick moss, grass and leaves is as good as anything nature can provide, but in temporary overcoats she can produce something immensely superior in a layer of snow. Freshly fallen snow contains so much air—a foot of snow melts down to only an inch of rain—that it is an extremely poor conductor of heat and blankets the ground from the cold of the snow surface. Snow is also a very efficient reflector of the sun's radiation during the day, and it disobeys the law that a good reflector is a poor radiator as it is an excellent radiator at night. Hence we find abnormally low temperatures above the snow on calm clear nights and falls to 0° are fairly common; but thanks to the poor conductivity of snow this drop in temperature is not transmitted through the snow to the ground beneath and even after four days and nights when the air temperature over the snow averaged 19°, the surface of the soil under the snow was found unfrozen. The details of this remarkable frost with a snow cover are given in Table 7, and it will be seen that as the soil surface under the snow was still unfrozen at 6 a.m. on November 14th when the air above the snow was 5°, it is obvious that at all other times the

NATURE'S WAY

bottom of the layer of snow must have been thawing, which explains the gradual reduction in the depth of the snow from 6

TABLE 7

Some Temperatures after a Snowstorm

Date and hour		Air over snow	Soil Surface under snow	4-inch Depth	8-inch Depth	Depth of snow lying
	1919	deg.	deg.	deg.	deg.	
	6 a.m.	31·3	32	34	35·8	Snow ceased 6 p.m.
Nov.	noon	31·3	32	33·8	35·8	
12	6 p.m.	31·3	32	33·8	35·8	6 in. deep
	midnight	30·9	32	33·8	35·8	
	6 a.m.	18·0	32	33·8	35·8	
Nov.	noon	27·0	32	33·8	35·8	4 in. deep
13	6 p.m.	13·0	32	33·8	35·8	
	midnight	8·5	32	33·8	35·8	
	6 a.m.	5·0	32	33·8	35·8	
Nov.	noon	16·5	32	33·8	35·8	3 in. deep
14	6 p.m.	14·5	32	33·8	35·8	
	midnight	18·0	32	33·8	35·8	
	6 a.m.	8·5	32	33·6	35·6	
Nov.	noon	23·7	32	33·6	35·2	2¾ in. deep
15	6 p.m.	20·3	32	33·6	35·2	
	midnight	15·0	32	33·6	35·2	
	6 a.m.	19·5	32	33·6	35·0	
Nov.	noon	29·7	32	33·5	35·0	2¼ in. deep
16	6 p.m.	32·0	32	33·5	35·0	
	midnight	32·0	32	33·2	34·8	
Nov. 17	noon	44·0	32·7	32·9	34·5	Snow melted

inches to 2¼ inches although the temperature of the snow surface was never above freezing point until the thaw set in on the night of November 16th.

When there is no snow cover a layer of frozen soil acts in much the same way—though not so efficiently—as a layer of snow. At the beginning of the spell of frost, and before the surface actually becomes frozen, the fall in temperature underground is of course fairly rapid as it follows with normal amplitude and lag the fall of the surface temperature. But as soon as the surface freezes this rapid fall is replaced by a slow steady decrease in temperature which is in strong contrast to the former rapid fall. The much smaller changes in temperature underground during a spell of frost are due to the fact that when once the surface is frozen a layer of invariable temperature at the bottom of the frozen layer which is always at 32° lies between the surface and the lower depths. The surface may fall, like the snow surface, to very low temperatures but, owing to the latent heat liberated continually at the bottom of the frozen layer, the temperature gradient in the frozen soil is very much greater than in the unfrozen soil beneath, and the temperature variations are not conducted uniformly downwards, and the only result of low surface temperatures is to increase slowly the depth of the frozen layer.

But frozen soil is not so efficient a winter overcoat as snow, for a fall of snow may be sudden and fall on ground not yet reduced to near freezing point, while a spell of frost must at least reduce the surface to 32° before the blanket of frozen soil can appear to ward off further low temperatures from the ground beneath.

From the point of view of the gardener it is best to have snow first and the frost after that, when plants will come to no harm beneath a few inches of snow cover from extremely low temperatures which would do a lot of damage on open ground without snow. The farmer may lose his sheep in deep snow, but there is a good chance that his dog may locate them and the farmer may dig them out little the worse many days later. Elizabeth Woodcock, who in February 1799 was lost in the snow outside Cambridge, was found eight days later still alive. The householder looks to the snow to save his cisterns and pipes in the loft from frost, and the hedgehog—though he probably does not think about it at all—will emerge safely in the spring from his nest of leaves in the hedgerow if only the snow is deep enough.

Thus during the process of freezing and thawing until the whole frozen depth is thawed and the temperature gradient through the soil again becomes uniform, surface variations of temperature hardly affect the 4-inch depth thermometer at all.

During the frost of December 31st–January 4th the frost reached the 4-inch depth after about 100 hours during which the average surface temperature was 30°.

If thick snow falls on hard frozen ground in February, then when the thaw comes we are reminded of the country saying, February fill dyke, which is so often true in spite of the fact that February is a dry month and so, it is argued, cannot fill the dykes. February certainly is a dry month, in seventeen out of about a hundred years it was the driest month of all, April and June coming next with fourteen years. Three out of the ten most notable droughts in the last sixty years were in February, and February 1932 ranks as the driest month on record over the British Isles generally. Yes, there is no doubt it is a dry month, but the argument that therefore it cannot fill the dykes is fallacious; it is not the total amount of rain but what happens to it that settles whether the dykes fill or not, and when the heavy snow melts, all this water, since it cannot percolate into frozen ground, runs away into the dykes.

In my Derbyshire garden we had a rather steep gravel path which was drained every few feet by a cross gully of a half-tile drain to take the water to one side and prevent it from eroding the gravel. After hard frosts this gravel path became very soft and a lot of water ran out of it down the gullies during the thaw. One February we had a very hard frost which penetrated down about eight inches and then six inches of snow fell on the top of this frozen ground followed by a rapid thaw. As soon as the snow ceased we isolated a ten-foot length of the path between two gullies by clearing the snow all round it and we led the drain which would collect the water when the thaw came into a sump where it fell into a bucket.

The snow gauge when all the snow was melted gave $\frac{1}{2}$ inch of melted snow, but the buckets collected over 9 gallons of water which gave a total of $\frac{5}{8}$ inch over the collecting area. Of this, $\frac{1}{2}$ inch was the melted snow and the rest was given by the frozen soil before it was all thawed. Later on in the same month we had another hard frost without snow which on thawing gave $\frac{1}{12}$ inch of water over the whole collecting area. The total rainfall for the month was just under an inch, of which, as we have seen, over ⅔ inch ran away into the drain while we were keeping the records. So this February certainly filled the dykes well with a large proportion of its low rainfall.

VIII

Frost

Our home when we were young was built on one of those small hills, capped with Northampton sands and Great Oolite Limestone so typical of south Northamptonshire and was about seventy feet above the level of the Lias valleys which lay all around it and in which flowed the Tove and its tributaries. Our orchard on the top of the hill might suffer from wind but not from frost.

On a clear autumn evening the mechanism was obvious, the smoke from a bonfire in the orchard wound its way down on the ground from the hill-top to the valley of the Tove as a river of smoke, showing how the air that carried it, robbed of its warmth by radiation from leaf and twig and grass on the summit, spilt over the edge of the hill and was replaced by warmer air from aloft. On several nights after sunset when the river of smoke was rolling valleywards, I climbed the tall cherry tree in the orchard with a thermometer in my pocket and compared the temperatures high up in the tree and on the grass of the orchard. The greatest difference I ever got was 10°, but that was convincing proof, coupled with the river of smoke, that the air of the orchard was warmed by fresh supplies from up aloft while the cooled air flowed away down the hill on the ground.

By morning the hoar frost over everything in the meadow was a marvellous sight in the sunshine, while usually on the top of the hill there was plenty of dew but no hoar frost. Quite often the minimum in the meadow was 7° lower than in the orchard, and it took a spring frost of real severity to spoil our fruit blossom. Our friends at Potcote, a little village that lay in the hollow and ringed round by the heights of Cold Higham from which cold air poured upon them all night from all directions, complained bitterly that their dahlias were often killed by frost in August. And one of them, whose grounds were on the southern slope of the hill, only succeeded with a new orchard he had planted by building a V-shaped wall above his orchard to deflect the stream of cold air on the slope to either side of his orchard as it passed him. This and the fact that he kept the ground beneath his trees dug and weeded, so that it warmed up well in the sun by day, enabled him to have some fruit in this treacherous hollow, which from the name of Cold Higham must have been known as a cold spot since Domesday times.

So Potcote may claim to join those other frost hollows, which have the reputation of being the most exceptionally frosty places in the British Isles, and of which the Rickmansworth frost hollow is probably the most famous. Phenomenally low temperatures at all seasons of the year have been recorded there, and quite often the minima are up to 12° lower than in neighbouring districts. In this constricted valley in the Chilterns day temperatures are high and night temperatures low and on 29th August 1936, with a minimum of 34° and a maximum of 85°, it achieved a range of 51°, one of the greatest daily ranges recorded in England.

Another factor in the incidence of frost is the nature of the vegetation cover, and the lowest minima are generally found over long coarse grass. The grass blades are good radiators and become so cool that the air caught between them is chilled, and as the grass stems and roots are very poor conductors of warmth from the earth beneath there is nothing to counterbalance the loss by radiation. A town garden is generally warmer than a country garden in the same district. As soon as the sun begins to give warmth in the spring the roof and walls of the house absorb heat by day and give it back at night, and even in winter the walls of the houses give out some warmth conducted from the heated interior. It is seldom that snow will lie close to the wall of a house, and a bed close under the wall is often unfrozen after a hard frost in the open.

All of which suggests that forecasting frost is a very local

FROST

problem, and whether there will or will not be a frost at a given place on a calm clear night depends on many factors. We have seen that the site and the vegetation cover count for a great deal, and after these the kind of soil and its moisture content and conductivity, the surface and underground temperatures at sunset, the effect of humidity or radiation, and the dewpoint all come into the question. We may well wonder if the task of forecasting a local frost is one suited to an amateur; and we may not be surprised to find that many books on the weather give correctly the factors on which the chance of frost depend, yet later they ignore all these factors except the dewpoint, and suggest that the amateur should base his forecast on that. Presumably this is just to give him something easy to do, regardless of whether the forecast so made is likely to be of much use or not.

Some of the methods based on the dewpoint forecast the minimum temperature for the morning as either the dewpoint itself, or the dewpoint with from 4° to 8° subtracted from it. These are at once suspect by anyone who has taken hourly readings of temperature during the night. The normal temperature curve on a calm clear night falls very steeply for several hours after sunset, then the rate of fall gradually slows down, and at the dewpoint may almost cease, so that the temperature curve traces nearly a horizontal line for a time; but later the fall is continued and the final temperature at minimum may be a number of degrees below the dewpoint which varies from night to night and cannot be predicted. The shape of the temperature curve also disposes of the idea prevalent in America that the rate of fall from maximum to minimum is uniform and by doubling the fall half way through at about 10 p.m. the total fall can be computed. Had the fall in temperature at the surface in Figure 5 for the first half of the night been repeated in the second half the minimum forecast would have been 18° instead of the actual minimum of 33°.

As Figures 5 and 6 are the basis for all our forecasts they are worth a little study. They have been chosen as being fairly typical temperature curves for the surface and 4-inch depth when dew was deposited but no frost occurred and when the surface

FROST

was frozen with hoar frost, the temperature curve of the 4-inch depth has its maxima and minima about three hours later than those at the surface, so that the 4-inch depth curve has not quite reached its minimum at the time of surface minimum, and as a

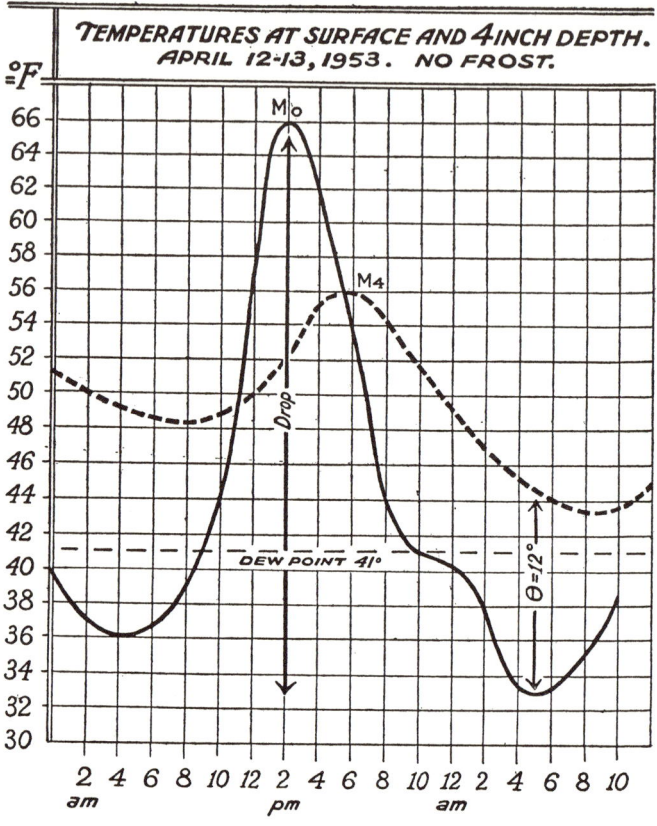

FIGURE 5

general rule has only completed about five-sixths of its range from maximum to minimum. As long as the surface curve is above the 4-inch depth curve heat is being accumulated underground to be returned to the surface to combat loss of heat by radiation when the surface curve falls below the 4-inch depth curve about 5 to 6 p.m.

FROST

After the maximum the surface temperature curve falls very steeply and continues to do so until the difference in temperature of the surface and 4-inch depth enables enough heat to be transmitted from the 4-inch depth to the surface to balance the outgoing radiation. As radiation is dependent on relative humidity this number of degrees can be calculated if we know the relative humidity of the night fairly accurately. The humidity on the night of April 12th–13th averaged 60 per cent and the number of degrees was 12°, while on night of April 3rd–4th it averaged 70 per cent and the number of degrees was 11°,

at 90% humidity the number of degrees = 9 to nearest degree
80% humidity the number of degrees = 10 to nearest degree
70% humidity the number of degrees = 11 to nearest degree
60% humidity the number of degrees = 12 to nearest degree
50% humidity the number of degrees = 13 to nearest degree

In Figure 5 we see the effect of latent heat liberated in the surface when dew was deposited there since the dewpoint was above freezing point, and in Figure 6 the same effect from the latent heat liberated when hoar frost was deposited on the surface, and the surface was frozen as the dewpoint was at the freezing point. It is these kinks that make any forecast based on the dewpoint only so unreliable.

For my own garden the drop in the surface temperature from the day's maximum to the following morning's minimum can be very simply estimated at about 6 p.m. by taking the difference of the surface and 4-inch depth maxima, and adding to it the appropriate number of degrees from the humidity table, and multiplying their sum by $1\frac{1}{2}$. Thus on 9th April 1953 the surface maximum was 64, the 4-inch depth maximum was 50, the humidity was 80 per cent and the appropriate number of degrees was 10. So the drop was $1\frac{1}{2}$ times $(14 + 10) = 36°$ and the forecast minimum was $64 - 36 = 28°$ which agreed with the actual minimum. Where soils are different to my sandy loam, the multiplier will have to be adjusted to make the forecasts fit the actual minima recorded. It will be found that instead of $1\frac{1}{2}$ a multiplyer lying between $1\frac{1}{3}$ and $1\frac{2}{3}$ may have to be used.

For those to whom my forecast equation appears too com-

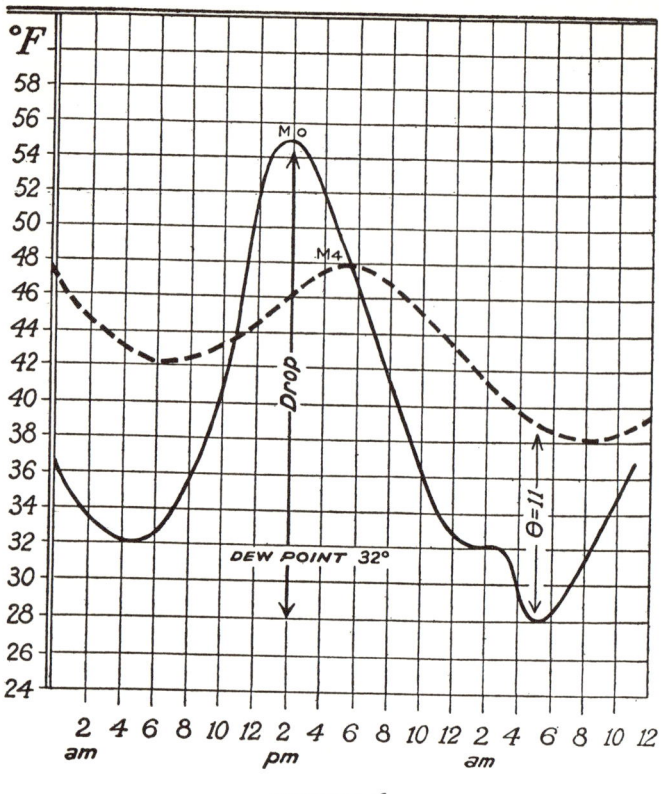

FIGURE 6

Temperatures at Surface and 4-inch Depth
April 3–4 1953. Frost

plicated a study of soil temperature records over many years suggests a short cut of reasonable accuracy, depending only on the maximum temperature at the 4-inch depth, and so available to anyone who has only a single maximum thermometer in action. If the maximum at the 4-inch depth in my garden is 53° or over there is little chance of a frost, but if the maximum is below 53° and the humidity is low, at least one degree of frost is likely for every degree below 53° that the 4-inch depth maximum falls. Using the temperatures given above for 9th April 1953 and this short cut method we should have estimated at least three degrees of frost and a minimum of 29° against the actual minimum of 28°.

When the surface did not freeze the forecasts were very accurate, when a frost actually took place a slight reduction of the forecast minimum below the actual minimum was inevitable, as to keep the calculations fairly simple the latent heat liberated has been neglected in our equation. But this slight error of a degree or so is on the right side and with frost it is better to be sure than sorry. The method has proved its usefulness in the orange and citrus groves of California, where friends have told me of the savings they have made by not putting in action all the frost protection measures when there was no need for them.

The forecast minimum temperature is of course worked out for my own soil and garden and ideal conditions of clear sky, no wind, and moderate humidity. A slight breeze, a temporary clouding of the sky, or a very heavy deposit of dew or hoar frost will prevent the surface falling to as low a temperature as was expected.

Years ago, when anyone enquired how many degrees of frost a particular tree or plant would stand we thought we knew the answer; but to-day, since we have found that the susceptibility of plants to frost is dependent on so many factors, we know that there is no unique answer to this question. The soil, the succession of cold and warm spells, the wind and humidity, the contained moisture and sugar content of the sap, and the differences between varieties of the same species all have to be considered before we can estimate any plant's behaviour under frost. Plant tissues commonly supercool to about 28° before any ice formation takes place, but a single crystal of hoar frost, when the dewpoint is below 32°, may act as a catalyst and turn the supercooled tissues into a frozen mass. On such small issues may the life or death of a plant depend.

For example, for many years I should have said that Christmas roses were immune to frost since I have known them survive many hard frosts with temperatures down to 12°. But now I know that it is not the hard frost that kills them, but the frequent

FROST

freezing and thawing in minor frosts of only a few degrees. So now after the first hard frost I cover them with cloches so that they escape the minor frosts; but for good flowers it does not pay to cover them too early, it is essential to let them endure a frost or two before giving them protection.

The hydrangeas, which the nurserymen pot up in small pots and force into flower in the greenhouse ready for sale in the shops about Easter, have had their natural rhythm disturbed by this treatment and are interesting to restore to hardness and growth outside in the garden. Normally they are shrubs with a high sugar content in their sap in the winter and so are not liable to coagulation of the protein of their cells which is the common cause of death by frost; but in their first summer out of doors they do not regain this normality and so have to be coddled.

I cut them down after flowering, repot into a much larger pot and bury the pot in the earth of my sheltered nursery bed which is protected by wind-breaks on every side but the south. There they grow new leaves, rather later than usual, which need hardening off slowly to stand a mild degree of frost. Having got them as far as this one might expect all would be well, but a spell of mild weather in winter soon destroys this first-earned hardiness, and they will die in the next frost if they are not given protection. In one garden, where I had an underground ice-house, I put them in there as soon as a mild spell started, and so prevented loss of hardiness and could put them outside as soon as it got colder again. But normally they must be given protection at intervals through their first winter, after that is over they appear to regain their natural rhythm and grow into big bushes which flower profusely. These few examples show how complex a question frost susceptibility may be, and how impossible it is to give an exact figure for a killing temperature without knowing all the relevant facts.

If our forecast suggests that a frost is imminent and we have some tender plants that must be protected, then we must put protection methods into action according to the severity of the expected frost, and whether the tender plants are bedded out or in pots and boxes and so can be moved. For those that cannot be

FROST

moved, bird netting has a maximum efficiency of a degree or two, and brown paper or sacking up to 5°. In my young days, when I rode about the countryside in the early hours of spring mornings, I saw many potato plots covered with brown paper in the cottage gardens, and quite a few tramps asleep under the hedgerows also wrapped up in sheets of brown paper. If the expected frost is likely to be too severe for these protections, we must fall back on our cloches with a maximum efficiency of 7° to 10°.

But if our plants are movable we can collect them in a shelter zone which will ward off 2° to 3° of frost, or put them under a south-facing wall with an efficiency of 5° to 7°, or in a cold frame which will protect up to 7° to 10° of frost, or under cloches along a south-facing wall which have in my garden given protection up to 12° to 17°.

That, I think, exhausts the resources of an average gardener; if we are unlucky enough to have a night when the temperature goes down to zero, nothing but a heated greenhouse or electrically heated frame will give enough protection. When fruit is grown on a commercial scale more elaborate frost protection methods are necessary. The old methods of smudge fires are now rather discredited, partly owing to their expense, their inefficiency, and the damage they caused to neighbouring properties. To-day one of the newest methods is to sprinkle water over fruit blossoms when the temperature falls to 32°, and so to warm the blossoms by the latent heat liberated by the water in turning to ice. As a single wetting gives only short protection a continuous fine spray is necessary. This is expensive owing to the high cost of installing the sprinkling apparatus. and the use of 2.000 gallons of water per acre, per hour. The heat given to the blossoms during freezing has, of course, to be given back when the ice is thawed, but this can be taken out of the rising temperature of the air in the morning and not taken from the tree. To the uninitiated the sight of an orchard of apple trees in full blossom covered with icicles is alarming, but frost damage is averted without any damage to the blossom or the trees.

Frost can be good for the garden when we have nothing

tender which it can destroy. Frosts in the early winter help to harden off the plants we hope to winter out, and the harder late frosts are wonderful improvers of a heavy soil. When we are in that happy frame of mind that no harm and much good may come of it we can amuse ourselves making an estimate of the depth to which any spell of frost will penetrate the ground. The first frost is generally light and only just freezes the surface because the 4-inch depth temperature was fairly high; but the second night, when this has dropped to nearly 32°, the rate of freezing is much faster.

If we multiply together the average number of degrees of frost and the number of hours it is maintained, divide by 12 and then take the square root, this is the number of inches to which the frost will penetrate.

Thus on 14th November 1919, after a spell of 18 hours when the average number of degrees of frost was 10 the frost did not quite reach the 4-inch depth thermometer when the calculated depth was 3·9 inches. From November 28th–1st December 1952 was a continual spell of frost for 96 hours with the average of 6° of frost and the frost rather more than reached a 6-inch depth thermometer when the calculated depth was just under 7 inches. A frost of small intensity but lasting a long time will penetrate deeply; after 100 hours when the average was 2° of frost the frost had penetrated beyond the 4-inch depth, but to get to this depth in a night of 12 hours takes an average of 16° of frost.

In the great frost of 1895 when the mean temperature in London from January 26th to February 19th was 26° the earth thermometer at a foot depth registered 28°, and the calculated depth frozen was 17 inches. But London clay is very different soil to light loam, and the calculation is not really applicable. For the calculation is only suited to soils similar to that in my garden. Those who have different soils and moisture contents will have to divide by a number slightly different from 12 to adjust the result to their own conditions.

IX

Air Temperatures

A range of 50° in air temperature on any day is rare, but a range of 30° in air, and of 50° in the soil surface and of 20° at the 4-inch depth is comparatively common. The air temperature is largely controlled by the heat of the soil and the raising of the temperature of a block of soil one foot square and 4 inches deep by an average of 35° probably requires about 350 British Thermal Units. A British Thermal Unit is the amount of heat required to raise one pound of water 1° Fahrenheit; our gas bills are reckoned in therms which are units of 100,000

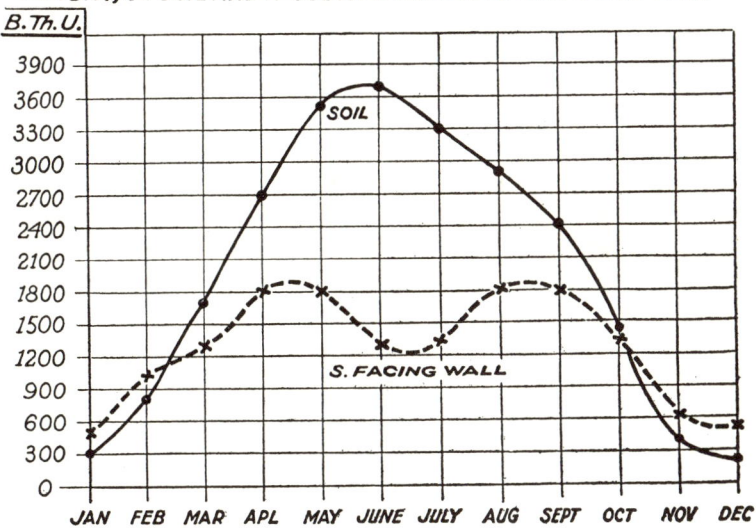

FIGURE 7

B.Th.U. But Figure 7 shows that the direct and diffused radiations from the sun on a soil surface on a fine summer day in June amounts to 1,800 B.Th.U. per square foot for the half day from minimum to maximum, so that only a little over a fifth of the sun's heat is used up in warming the surface soil. Of the rest some is lost by reflection and re-radiation, some is used to evaporate the moisture in the surface soil, some to warm the air and the depths of the soil below by conduction.

The grass lawn uses even less of the sun's heat in warming the turf. The leaves are good reflectors and excellent radiators, so that the range of air temperature over grass is less than over soil, and the temperature on the surface of the lawn is often 10° less than that of the surface of the soil, while at night the grass minimum is notoriously lower than the air minimum in the Stevenson Screen. It is a curious meteorological anomaly that the recording of frost is not done by the thermometer in the Stevenson Screen, nor by a thermometer at the surface of the soil, but by one placed with its bulb in contact with the tips of the grass blades on an area covered with short turf. This registers a frost if it falls to 30° or below.

This seems a very roundabout way of deciding whether there has been a frost or not. The thermometer in the screen obviously cannot do the job, for though it decides the minimum temperature that appears in the papers, this is so far removed from reality that we can read that the official minimum was not below freezing point, when the minimum in our own garden went down to 22° and the bathroom waste was frozen up. To the gardener it seems strange that the temperature on the grass—which apparently does not care whether it freezes or not, or whether freezing point is counted as 30° or 32°—should be used instead of the temperature on the soil where his plants do grow, and which really reckon 32° as freezing point and do not much care for a temperature of 30°.

No doubt those who wanted to arrange a standard exposure for the thermometer hoped to get a better standard on grass rather than on soil which might be sand, loam, or clay. But nature abhors uniformity, and the grass on which our thermo-

AIR TEMPERATURES

meters are exposed may be the short turf desired, or thick mossy turf, or very old turf whose matted roots go down many inches and insulate the top grass layer from all contact with the underground warmth. The different temperatures recorded on these different kinds of turf are very surprising, and everywhere, except inside the Stevenson Screen, nature is capable of defeating the best efforts of all who want a standard exposure.

In my Derbyshire garden I kept up a weather station complete with Stevenson Screen, sunshine recorder, anemometer and all the usual thermometers and hygrometers. So I was able to find the connections for that soil and situation between the readings of the Screen minimum, grass minimum, air over soil minimum and soil surface minimum on a hundred nights when frosts were likely.

On about half the nights the air over soil minimum was 2° higher, soil surface minimum was 4° higher, and screen minimum was 7·5° higher than the grass minimum. The greatest difference between the Screen and grass minima was 12° and the least 4°. So as a general rule when the grass minimum was 30° and recorded a frost, the air over the soil was 32°, the soil surface was 34°, and the screen temperature was 37·5°. In my Derbyshire garden the air over the soil gave a straightforward record of frost.

The tarmac road surfaces of the run into the garage or perhaps of a garden path are very good utilizers of the sun's heat on a summer day, and the layer of air at the surface of the tarmac may be up to 40° hotter than the air a couple of feet above it. So the difference in density of the two layers is so great that the rays of light are bent upwards as if there was a mirror on the ground, and the sky can be seen reflected in the road or path surface giving the illusion of pools of water on the tarmac. On occasion the refraction is so perfect that only the upper half of people approaching is visible and they appear to float in a shimmering lake of water.

Another very efficient trap for the sun's radiation is a south-facing brick or stone wall. The lower curve of Figure 7 gives the direct and diffused solar radiation on such a wall through the

AIR TEMPERATURES

year, and the maxima in spring and autumn give an extremely useful addition of temperature to ward off late spring frosts and to ripen the fruit in the autumn. It is difficult to give exactly the fraction of this total radiation absorbed which is given back to the air close to the wall or to the border under the wall, but the temperature values given in Table 8 show how much warmer it can be close to the wall than six feet away from it during the spring and summer months of 1954.

TABLE 8

Mean Temperatures under South-Facing Wall and in Open

Date	Air under Wall	Air under Cloche under Wall	Air in open
1954 February	degrees 39	degrees 45	degrees 35
March	45	54	39
April	56	66	47
May	61	71	51
June	68	78	59

In April and May when the radiation absorbed by the south-facing wall was at its maximum the air under the wall was 9° and 10° warmer than in the open, while the air under a cloche under the wall was 19° or 20° warmer than in the open. This shows very clearly the advantages of a walled garden, and what chances there are of tender crops coming to maturity with the combination of a cloche and a warm wall. In winter the advantages are apparent; after a sunny day if the following night is calm and clear, the frost line is clearly marked out several feet from the wall, and when snow comes the first light snowfall refuses to lie inside about the same distance from the wall until the heat of the wall has been dissipated.

The Benedictine and Cistercian monks were probably the first to realize the increase of warmth on and under walls, and in their gardens judged the efficiency of a wall by the speed of

5. Dewdrops on twigs and berries after a calm clear night

6. A dew pond on the South Downs

germination and ripening of their crops at different distances from it. Besides the ordinary wall fruits of apricots, peaches, and nectarines, they grew grapes, and winedressers are frequently mentioned in the abbey chronicles as part of the normal staff of a monastic house. William of Malmesbury, writing about 1150, tells us that the vale of Gloucester 'exhibits a greater number of vineyards than any county in England, yielding abundant crops and of superior quality: nor are the wines made here by any means harsh or ungrateful to the palate, for in point of sweetness they may bear comparison with the growths of France'.

That grape growing practically died out later may well be due to the dissolution of the monasteries and the dispersal of those who knew the art of wine making, rather than to any supposed decrease of summer temperature about that date. We have the parallel case where the art of ley farming, so assiduously practised by the monks in Scotland, died out completely for about 150 years after the dissolution of the monasteries, before it was revived about 1750. In 1786 F. X. Vispré, in his *Dissertation on the Growth of Wine in England* stated categorically that any part of England that would ripen wheat in August would ripen grapes in September. And this seems good common sense, because we have seen that a difference of about 4° below the temperature of the wine making areas of the north of France—which is claimed to make the growing of grapes out of doors in England impossible—can easily be more than made up in walled gardens and probably in many sheltered areas.

The Elizabethans grew new flowers and vegetables under the warm brick walls of their houses and ripened grapes, apricots and peaches in their walled gardens which at that time were all the fashion. Gordon Manley, in his book *Climate and the British Scene*, mentions a Jacobean walled garden where peaches are still grown in Teesdale, and two plum trees under a south wall near Alston in Cumberland which still ripen fruit 1,100 feet above sea level. In the walled garden of our old home in Northamptonshire, besides the peaches and nectarines on the walls, there was one particular walled arbour, which faced the south and had flanking protecting walls on the east and west, where we grew apricots, figs and a vine. The two first ripened their fruit every summer, but the vine only in the hottest years.

One of the most impressive features of our climate is the opportunity it gives by small changes of location, aspect, shelter, soil and method, to develop micro-climates over small areas where unusual crops can be grown quite successfully. *Vineyards in England,* edited by Edward Hyams, gives a detailed account of every aspect of the new viniculture which has been started in Surrey and Kent; but there is still plenty of scope for the amateur to try to grow the ordinary fruits in latitudes and at heights not previously tried, or to attempt to grow the olive—which so far has refused to ripen on the Cornish Riviera—or the pomegranate which occasionally will ripen in the south of England.

So far we have considered how micro-climates and environments affect the crops, now we must see how the crops affect the micro-climates. Up to now we have been accustomed to think of the soil surface as the hottest place by day and the coldest by night. But the thermal efficiency of the rabbit's form and the covering of moss and grass has shown that with a growing crop the warmest place by day and the coldest by night tends to be near the level of the top of the densest vegetation and that this active surface rises as the crop grows. The gradual change of the temperature gradient for the bare soil and a well-grown crop is plainly shown in Figure 8. On the bare soil the hottest point by day is the surface, with a steep drop in temperature as we rise above the surface or go down below it. With a crop the hottest point is generally about the level of the top of the crop—mustard at 2 inches, clover at 3 inches, stocks and asters at 4 inches, and rye-grass at 6 inches. The close canopy of the leaves, which during the day have been a good absorber of the sun's radiation, turns after sunset into an efficient radiator and quickly loses its heat to a clear sky and so becomes the coldest place at night.

So long as the plants give a complete cover to the soil they appear to reduce the steepness of the temperature gradient in comparison with bare soil, to change the height at which the

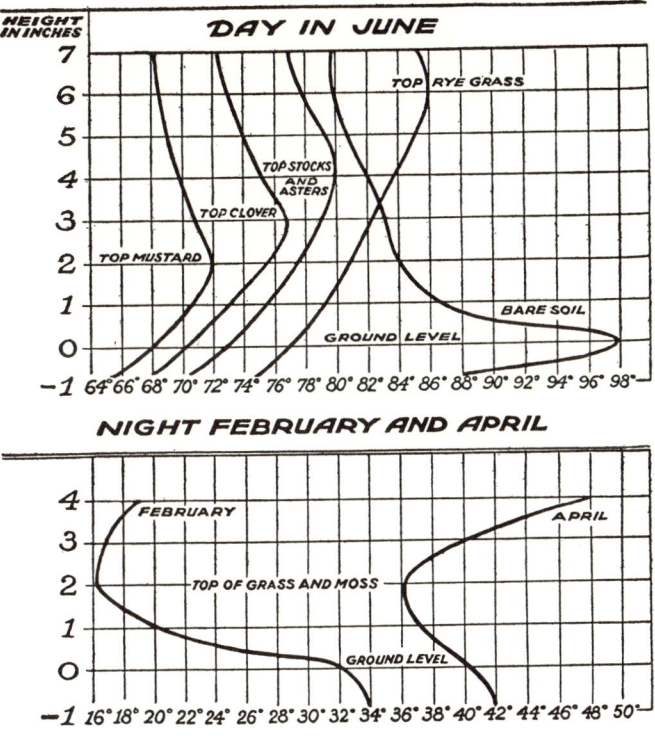

FIGURE 8

greatest heat and cold occurs, and to reduce the amplitude of the temperature variations. But if the plants give only a sparse cover the height of the maximum temperature is lowered, in a potato crop to about two-thirds of the way up the haulms, in a hay crop to about three-quarters of the way up the tallest grass heads, and in a cornfield to about half way up the crop. Most shrubberies behave similarly depending on the thickness of the bushes, as do raspberry canes, but dense woodland gives much the same results as beds of plants growing close together and it has its highest temperature among the tree tops, which accounts for the deliciously cool feeling in a wood on a very hot day, and for the warmth of an evergreen wood in the morning after the

AIR TEMPERATURES

first frost of the autumn when the coldest area in the tree-tops may be 25° colder than the soil surface.

Just as the soil only uses a small proportion of the sun's radiation in heating itself, so the canopies of vegetation use most of the sun's radiation in evaporating the transpiration water and not in heating themselves, and so avoid the death that would ensue if all the energy of the sunlight was used in heating and promoting chemical changes in the leaves. A sunflower leaf uses only about 8 per cent of the bright sunlight in warming itself and the other 92 per cent in evaporating the transpiration water; this explains why the rise in temperature of the vegetation is so much less than that of the soil which uses about 20 per cent of the bright sunshine in heating its surface layers.

Insects, being cold blooded, are very sensitive to high temperatures. Ants are much more active and move faster when the day is very hot, and crickets are supposed to give an accurate estimate of the temperature by the number of their chirps in thirteen seconds, with forty added. Many insects react to different temperatures over open soil and amongst vegetation. As a boy I used grasshoppers as bait for fishing and knew that the best way to catch them was to go to the lucerne field on a sunny day with a butterfly net. Every time the sun went behind a cloud the grasshoppers hid in the thick foliage of the lucerne, but as soon as the sun came out and the top of the thick foliage became uncomfortably hot they crawled up the flowering stems of the lucerne well above the hottest part and there they were easily caught with the butterfly net. Moreover, the tempo of the grasshoppers' stridulation depends on the temperature. In sunny weather they fiddle away merrily, but when the sun goes in their execution is slow and mournful.

In my Derbyshire garden I had a watercress pool fed by a spring which entered it by a narrow stone conduit, which in course of the years had become overgrown by a moss cushion which dipped into the water but was anchored in the stone on each side. One very hot June day, when I was examining the cushion, I noticed some tiny beetles sitting on its surface. When disturbed they burrowed into the moss but very soon reappeared

AIR TEMPERATURES

on the surface. This happened every hot cloudless day between 11 a.m. and 3 p.m. when I found the temperature of the interior of the moss was 82°, but the surface, due to evaporation of moisture, was only 70°. On days when the sun was in and out of cloud at intervals the activity of the beetles was intense and they came out and in exactly in agreement with the sun; on overcast days they were never in evidence at all.

Edwin Nörgaard, in his paper 'Biology of Danish Spiders', tells of a little black spider which makes its web between sparse plants of heather on stone fields and has to climb up its web to about two inches above ground level on hot days as the temperature there is often about 40° less than on the stony ground. And another spider that makes a house of small pieces of stone suspended above the ground, has to leave its home on hot days and live in the open where the air is cooler. If kept in its home during the heat of the day it dies.

X
Humidity and Dew

The beauty and wonder of dew still wakes in us much the same feelings as it did in our ancestors. Because a calm clear night is essential for its formation, the moon shines brightly as the dew forms, and so our ancestors supposed it fell from heaven and condensed out of the moonbeams. Early rose growers asserted that the beauty and perfume of the rose was due to the moon, and not, as we prefer to think, to the dew.

For many centuries dew was a beautiful and magical phenomenon; in the countryman's lore a portent of fine weather, to the girls of the village something to wash in to make them beautiful, to the more materially minded something to be used to rot the vegetable matter round the fibres by the dew-ret in the preparation of flax and hemp. Because the causes of dew are not simple to explain to the unscientific mind, dew still retains its magic; on a summer evening one moment the grass is dry and a few moments later it is wet with dew, something visible has been born from something invisible, and in the morning the roses will be strewn with diamonds and the grass roped with pearls.

In our youth we had many and various weather forecasting aids; the seaweed brought home from our last visit to the sea, immortalized in the music hall song of the period which began 'As soon as I touched my seaweed I knew it was going to be wet'; the long stout string with a heavy weight at one end which hung against the wall of the outer hall, and by a pin through the string gave a reading on a cardboard scale; Darby and Joan at the door of their cottage, he in a straw hat and with a walking stick came out when it was fine, she in a waterproof and with an umbrella, took his place when it was wet.

HUMIDITY AND DEW

Then we had a number of nature's aids, the scarlet pimpernel, the daisy, the chickweed, all of which closed their flowers before a storm; the everlasting flowers whose lower petals loosen and the flowers open when it is fine, but close up and shut the flower tightly when rain is coming; and those attractive coloured postcards with a charming view surmounted by a wide expanse of sky which in fine weather was a lovely mediterranean blue, but became red and lowering on the approach of bad weather.

It was not until we went to school that we realized that all our weather forecasters were really humidity measurers and performed their functions by virtue of the fact that the humidity generally increased a good deal before a rain storm. We learnt, too, that the coloured postcards with the variable skies were painted with cobalt chloride or cobalt thiocyanate which turns pink as the humidity increases.

I still use papers soaked in solutions of these chemicals to find the humidity in very small places where the ordinary more cumbersome humidity measuring instruments will not go. By keeping a set of standard tints to check each coloured paper by, it is possible to estimate the humidity within about 3 per cent accuracy from a piece of paper about the size of a little fingernail. The only disadvantage of this method is the rather long exposure necessary, so that it is not possible to get a result quickly as with a hair hygrometer or wet and dry bulb instrument. Cobalt chloride changes colour at about 50 per cent relative humidity and cobalt thiocyanate at about 75 per cent; seaweed feels moist at about 70 per cent, the scarlet pimpernel, daisy and chickweed shut up about 80 per cent, while Darby and Joan, the long string, and the everlasting flowers are more or less continuous operators over a scale which can be made from observations with the more scientific instruments.

What all these weather aids were measuring was, of course, the actual amount of water vapour in the air in comparison with the amount it could hold at the same temperature if it was saturated. So long as air at any temperature is not saturated it can take up moisture and evaporation will take place from the soil, vegetation, and the weekly wash. Only when the air is saturated, and there is probably a thick fog, does drying cease. Also hot air can contain a great deal more moisture than cold air before it becomes saturated, so that we dry our clothes before a fire or in the sun.

On the evening of a hot day as the air cools it gradually becomes able to hold less and less moisture before being saturated, and if the air falls low enough in temperature it must reach the stage when it is saturated. After that any further fall in temperature will mean that the moisture the air cannot hold must be deposited as dew or hoar frost according as the saturation temperature was above or below freezing-point. Just as we have to supply heat to vaporize water or melt ice, so water vapour has to give back this heat when it condenses into dew or becomes hoar frost. So by day the soil loses a good deal of heat when it evaporates the water in the surface, and may get some of it back at night from the dew or hoar frost deposited on the surface. We can get a faint idea of the heat lost by evaporation and gained from dew, by remembering how cold we felt when we fell into water and had to run home in our wet clothes in a wind, or how scalded we were when the lid fell off the kettle and the steam from the boiling water condensed on our hand.

High humidity has its advantages; as we have seen in Chapter VIII it makes hard frosts less likely on calm clear nights, for though the water vapour in the air absorbs very little of the incoming radiations of the sun during the day, yet it is an excellent absorber of the outgoing long wave earth radiation at night and acts as a greenhouse to trap the outgoing radiation. The increase of humidity as the temperature falls in the evening of a hot day is a boon to newly planted crops that probably have flagged during the heat of the day when temperature was high, evaporation active, and humidity was low. A nightly watering helps to restore their stiffness ready to face the next day's heat. The daily rhythm of high temperature and low humidity followed by lower temperature and higher humidity is desirable; it is only when high humidity becomes continuous over several weeks that it becomes a danger.

The growth of moulds and mildews does not begin until the

HUMIDITY AND DEW

relative humidity has been above 65 per cent for some considerable time, then if the humidity rises above 75 per cent it increases rapidly. And here it is important to note that damp wood or vegetation in an unventilated place will saturate the air around it and provide an excellent environment for moulds and mildews, even when the humidity of the ordinary open air is well below the critical value. It is only with difficulty that we can produce lower humidities than the normal free air, but at least we can ensure that we bring normal humidity into all places, that otherwise might easily become saturated and a source of mould and mildew, by letting what the gardener calls 'light and air' into them.

The way in which dew was formed and the source of its moisture was for long years a matter of controversy. Aristotle was more nearly right than any other for 900 years, for he defined dew as 'the humidity detached in minute particles from the chill clear air' while all the rest believed that dew fell from heaven. Dr. W. C. Wells, in his *Essay on Dew* of 1814, said dew was the moisture found on the surface of the earth during the night and early morning, particularly after a hot day, and produced by condensation of the vapour of the atmosphere on contact with the earth cooled by radiation. In 1885 Dr. John Aitken, in his paper, 'On Dew', in the Transactions of the Royal Society of Edinburgh, was even more specific and showed by experiment that the greater part of the dew was formed out of water vapour risen out of the earth to the surface of plant leaves. On its way it was trapped by the surface of the soil or by the grass or plant leaves which had cooled by radiation sufficiently to reduce the vapour below its saturation temperature. Aitken also gave some experiments whereby the amount of water lost by a piece of turf could be calculated, and on the assumption that all this vapour was condensed showed that about three-quarters of an ounce of dew per square foot was often found, corresponding to about one-hundredth of an inch of rain during a calm clear night.

The invention of plastics has made a measure of the dew deposited more easy than it was in Aitken's time. Many plastic

HUMIDITY AND DEW

materials are very light and a sheet of area one square foot weighs only a few ounces, and it has the very desirable property that dew deposited on it tends to collect in a multitude of small drops that do not readily run into each other and drop to the ground. Such a sheet supported on four short wooden pegs over the grass and put out about 5 p.m. on a spring or autumn night will have collected quite a lot of dew on its underside by 11 p.m. The sheet can be lifted and turned over without losing any of the dew, and can be brought in and weighed. As the dry sheet is so light even a quarter of an ounce of dew on it is readily detectable on a sensitive balance. After weighing, the sheet is dried and returned to its place to catch more dew until about 5 a.m., when the final weighing is done before the rising sun tends to evaporate any of the trapped dew. In this way I have collected about an ounce of dew—corresponding to over a hundredth of an inch of rain—in the south of England and rather less in my gardens in Derbyshire and Edinburgh.

In the four years I worked at Eastbourne I spent a good deal of time on Beachy Head between Eastbourne and Birling Gap. There on the downs I talked to the shepherds who watered their flocks of sheep at the dew ponds in the early morning, and they all agreed that, while the ponds at the foot of the downs often dried up in the summer, there was plenty of water for their sheep in the dew ponds on the tops. The ponds were of peculiar construction and, according to the shepherds, could only be properly made by an expert. The bottom was puddled chalk, then came a layer of straw to insulate the layers above from the warmth of the earth, then another layer of puddled chalk or clay. On the top of all this came a layer of chalk and flint rubble, and finally a concrete cover was made over the rubble so that the feet of the sheep could do no harm.

The ponds varied in performance according to their siting, and the shepherds said that the best pond on those downs lay in a gully on the northern slope just below the crest. It did not get so warm in the sun as the others which were all on the top, and it began to cool on calm clear nights before the sun had set. The air which cooled on the top continually flowed down the gully

HUMIDITY AND DEW

and over the dew pond which radiated its warmth from the concrete surface to the open sky and soon fell below the dew point of the flowing air, so that considerable quantities of dew condensed into the pond on good nights.

Many years later I came across a similar phenomenon on an asphalt playground on a slight slope below the crest of a hill in Derbyshire. Every summer morning about 4 a.m. after a calm clear night puddles of water could be seen on the asphalt. A neighbour of mine, who was interested in dew ponds, made a model pond on an ideal site on the north slope of his orchard, just below the crest of the hill well away from the fruit trees. He dug out a shallow saucer-shaped hole and lined it with straw, and fitted into it a black-japanned metal bowl of just over $9\frac{1}{2}$ inches radius so that the total catchment area was about 2 square feet. A hole in the centre of the bowl led any collected dew to a collecting jar graduated in fluid ounces, and every hundredth of an inch of dew deposited over that area gave about $1\frac{1}{2}$ ounces of water in the graduated jar, which had not to be read in the early hours of the morning as it was shielded from the sun by the crest of the hill, and was sheltered against evaporation by the metal bowl. Every suitable night some water collected in the graduated jar and the greatest amount of any night in one year was $1\frac{2}{3}$ ounces of water, corresponding to rather over a hundredth of an inch of dew. No record was kept of the amount of dew when the nearby rain gauge recorded any rainfall during the night.

The following year we changed the site to the top of the crest with the result that very little dew was recorded on most nights. It seemed that the site was most important; on the top of the crest the cold air that trickled down the slope drew a supply of warmer air from up aloft into the bowl of the dew pond, but this was never displaced as it could not drain away; but on the slope the continual river of cooler air that flowed through the bowl of the dew pond provided a source of dew all through the night.

Methods of collecting dew are in use in other parts of the world based on a very old method used in prehistoric times. In the deserts of Algeria, where the yearly rainfall is about a quarter of an inch, in spite of the low relative humidity at midday, the range of temperature is so great that by night the air is saturated and dew forms on pyramids of stone specially built to catch it. A pyramid with a square base of 30 feet each way caught four pints of water every night in the summer and a little every night in the winter, so that the amount of dew deposited in the year was about equal to the rainfall.

The yearly total of dew deposited in the south of England appears to be about $1\frac{1}{2}$ inches, but this is much exceeded in a country like Palestine where no rain falls in the summer from June to September and growth during that period depends almost entirely on the dew. The value of it is reflected in the many references to it in the Bible where it was always regarded as a blessing. 'As the dew of Hermon and as the dew that descended upon the mountains of Zion: for there the Lord commanded the blessing, even life for evermore.'

With the aid of several sheets of plastic we can test the amount of dew formed in several places in the garden. A sheet over bare soil will pick up nearly as much as over the grass, those on the crazy paving and rockery will have streaks of dew on them only where the gaps in the stones were, the one on the gravel path will be a mosaic of wet and dry patches corresponding exactly to the gaps and stones of the foundations of the path. The results of one good clear night in spring or autumn will give us convincing evidence that moisture rises out of the ground so long as the lower depths of the soil are hotter than the surface and will condense on our plastic sheet wherever it can escape through the surface. Wherever there was a bare patch on the plastic sheet we shall find something underground that has already trapped the moisture and has condensed it on its under side.

Next morning, if we watch our sheets, we shall see the dew is re-evaporated and the moisture driven back into the ground. In the same way the moisture on the underside of the stones in the crazy paving, rockery, or gravel path has gone back under ground as soon as the temperature of the surface has risen above that of the underground layers. So rockeries, crazy paving, and stony beds seldom suffer from lack of water in droughts, because they only lose a fraction of the moisture loss over open soil. Even the clods in the farmers' fields left from the autumn ploughing serve the same purpose, and wheat is sown in a knobby seed bed to help conserve the moisture in the spring. It is well known that implements and pieces of timber, left lying on the ground in very dry districts in India and tropical countries, encourage a growth of luxuriant weedy vegetation in the season of no rain from the dew they trap on their undersides during the night, and from the reduced evaporation under them during the day. It is always useful to know that we have in our hands such a simple remedy for drought which we can use in some special nursery bed when necessary.

The wood louse knows all about humidity and dew. It is a crustacean, and though it has become adapted to terrestrial life, it still finds damp necessary for its existence. Hence it is nocturnal in its habits and it spends the day under some slate or stone or piece of rotten wood where the humidity is very high, due to the condensation of dew on fine nights and the total lack of evaporation. So long as the humidity is high the wood louse will live in comfort under a piece of slate that is too hot to touch in the sun, but would die quickly if the slate was removed and it could not quickly find some other damp shelter.

Working in the micro-climate of the sheep's fleece in 1947, Davies found that only rarely did the relative humidity in the fleece exceed the value at which development of the larvae of the blow-fly become possible at the blood heat temperature of the sheep. The next step was the study of the conditions which led to sweating and the production of high local relative humidities; if these could be prevented then only rarely could the strike of sheep by blow-flies become a menace. Fifty years ago our old shepherd always tested his bunch of seaweed at his cottage door as he went out on a summer day; if he found it damp he knew it was the sort of weather he must be more than usually keen on his look out for struck sheep. It seems now that his idea was well founded.

XI

Wind and Shelter

Wind can do a lot of damage in a garden, and the gales of summer when the lupins and delphiniums are at their best seem to be annual events. But the less spectacular but more insidious destruction comes from persistent winds in dry weather or when the frost has penetrated deeply into the ground. It is always very difficult to find the reason for the casualties that happen in a garden, both in summer amongst the newly planted shallow-rooted bedding plants, and in winter amongst rose bushes, viburnums, daphne, and wintered antirrhinums. But by keeping a careful inquest on losses over many years it seems that a very large proportion of these losses can be put down to wind when the soil is very dry or hard frozen, and not to the drought or the cold as many people imagine.

Shallow-rooted plants, and newly planted trees whose roots have been pruned by the nurserymen before sending out, reach down to so small a depth in the soil that they are the worst sufferers. It is the top few inches that get dried out so quickly in a drought and the first six inches that get frozen most frequently, and when this happens the roots can supply no moisture to the stems and leaves to balance the losses from evaporation in a strong wind of low relative humidity. So the plants wilt and the stems harden and die back more and more as the drought or frost continues and the wind goes on, and eventually the plant or bush dies from want of water.

I have saved the lives of many rather special summer plants by watering them and then laying round them a mulch of compost held down by a few stones which help to conserve the

WIND AND SHELTER

moisture and prevent the birds from scratching away the compost. A square foot of soil open to the sun and wind lost three pounds of water in fifteen consecutive dry days; a similar square foot sheltered from the wind and covered with compost and stones lost only five ounces of water in the same period. Roses on a fence or wall can be covered by a piece of sacking, such as nurserymen wrap them in when they deliver them, to protect them against wind and frost; single bush roses in beds are more troublesome as most simple wind-breaks are liable to be carried away in a strong wind, but a bag drawn over four stout stakes round the bush is effective and will stand up to a very strong wind. In general it is only the new purchases which are liable to become casualties, the old-established bushes have probably pushed down their roots below the reach of any ordinary frost, and so can go on supplying moisture to their stems in frosty weather.

After losing most of my wintered antirrhinums each spring for many years I realized that those plants that seeded themselves successfully year after year grew mostly in sheltered places under the walls of terraces or in the corners of walled gardens. So now I transplant all mine into a corner of the garden between two fences and fix a sack across the corner which I can raise or lower at will. In wintry weather the sack is kept down as long as the frosts and winds last, and now I have no losses. This rough and ready shelter was very effective; on frosty nights the temperature under it averaged 6° higher than outside, while on windy days the shelter was 4° higher than outside, and the humidity was much higher. This higher humidity was an enormous advantage as will be seen from Table 9, which gives the evaporation rate at various wind speeds and humidities compared with that of still air at 95 per cent humidity.

These evaporation rates were based on the actual times taken to dry a cotton cloth, at the various wind speeds and relative humidities shown.

Last year we had a splendid example of what a long period of really cold winds and frosts can do to newly planted shrubs when no shelter was possible. Owing to various delays a hedge

WIND AND SHELTER

TABLE 9

Evaporation at Wind Speeds miles per hour

Relative Humidity per cent	Evaporation at wind speeds miles per hour				
	0	4–6	7–9	10–16	17–23
95	1	1	2	3	4
90	1	2	4	6	8
75	1½	6	7½	10	15
65	2	8	10	15	20

of laurels was planted in the middle of January 1953, and almost at once a spell of fourteen dry days of frost, high winds, and no rain followed their planting. After that we had a few mild days, and then another spell of twenty days very cold weather with more frost and high winds and still no rain. By the middle of March, thanks to combined frost, drought and wind, many of the laurels had lost most of their leaves and their branches had died back to half their original length. A spell of milder weather and copious watering saved their lives, but they were a miserable sight all summer until the new leaves came in August. But to-day, eighteen months later, they are beginning to make quite a good hedge and only two of them still show any signs of the ordeal they went through.

On our mountains and moorlands we can see the result of almost perpetual wind action in the shape of the trees and bushes. On the windward side the buds dry up and the stems die back, and only a few develop on the lee side, where they are already sheltered by the bent-over crown. So the trees are stunted and shaped 'like the flame of a candle in a draught', as Richard Jefferies so aptly put it.

These temporary shelters were all right in their way, but obviously the garden needed a sheltered spot where plants could winter in the open, and young plants could be grown in the spring in a favourable seed bed. So I set to work to discover which of the various sites would make the best open-air sheltered

7. On windy sites branches grow down wind

8. Even in winter trees give shelter from radiation

WIND AND SHELTER

corner, where wind would be screened off, frost could hardly enter, and yet the sun would shine into it freely.

Scotland has always been a windy country and for many years was almost destitute of trees. So bare of shelter were most of the farms, that the Parliament of 1457 ordained that all freeholders, temporal and spiritual, were to plant on their estates trees, hedges, and broom parks to improve the aspect of the country, and provide shelter. Acting on these instructions the Abbey of Coupar made a similar injunction in all their leases, and in 1473 the new tenant of Aberbothry undertook to plant on either side of his farm buildings a broom park, and his neighbours entered into a similar covenant. Broom was very suitable to wet soil and strong winds; it gave good shelter, was useful for fodder, was decorative and grew to a good height. After the dissolution of the monasteries the planting of shelter belts was probably discontinued, but the Improvers began it again in 1723, and Sir Archibald Grant of Monymusk had 67,000 young trees growing in his nursery, and had already planted or sold to his tenants over a million trees of all kinds by 1733.

The system of shelter belts they provided show that about a quarter of a mile between successive belts was considered a useful distance. As probably the belts were about 60 feet high, they apparently provided protection up to a distance of about twenty-two times their height, with the most complete protection up to ten times their height, as shown by the belts about 200 yards apart on either side of the farm steadings. These ratios must have been found empirically, and apparently are true irrespective of the height of the wind-break and the strength of the wind, and the protection extends up to about the level of the top of the wind-break. In reckoning the reduction of wind speed produced by a wind-break it is wise to remember that the official wind speeds given by the Air Ministry are for heights of thirty-three feet above the ground. At ground level the speed is probably about half the official speed.

Eventually I chose a corner in between two solid wooden fences on the north and east which gave complete protection from these directions. The corner was open to the west and

WIND AND SHELTER

south, and my wind-breaks were sited on the west side. They were four feet high and about thirty feet apart, the outermost a solid wooden screen, and the other a trellis on which grew two evergreen honeysuckle which gave good cover. The result exceeded my expectations and provided a zone free of wind, having almost complete protection from ordinary frosts and plenty of sun and air. The reduction of wind increased the humidity and so reduced the evaporation and transpiration with their consequent loss of heat to the ground and the plants. It also conserved the heat given off by the fences in the zone, and the presence of the two wooden fences and the trellis reduced the amount of sky open to the zone's radiation at night to a minimum. Both by day and night the zone was warmer than the rest of the garden as can be seen from the figures in Table 10.

TABLE 10

Temperatures in and outside a Wind-break Zone

Date 1954	Wind-break Zone			Outside			
	Highest Max.	Lowest Min.	Mean Temp.	Highest Max.	Lowest Min.	Mean Temp.	Diff. of Mean
January	degrees 55	degrees 28	degrees 40	degrees 52	degrees 22	degrees 37	degrees 3
February	55	25	38	50	14	35	3
March	58	26	42	54	21	39	3
April	78	32	52	70	27	47	5
May	79	33	58	70	30	51	7

The success of so simple a pair of wind-breaks suggested that even less elaborate screens might be reasonably effective. Sir A. D. Hall in his book *The Soil* gives figures for an April day with a strong N.E. wind when a slight hedge of spruce boughs about two feet high kept the temperature of a plot 2° to 3° higher than that of nearby open ground.

So I made two different enclosures; the first was a six-foot square of wooden trellis, four feet high, with six-inch square openings; the other a cylinder of wire netting of about four feet diameter and made of rabbit netting four feet high. These were tested on 1st August 1953, a fine sunny day with a nice breeze, just enough to test the wind-break action of both enclosures and the resultant effect on the relative humidity and temperature inside them. The results of the test are given in Figure 9 where it will be seen that the trellis cage was a much more efficient wind-

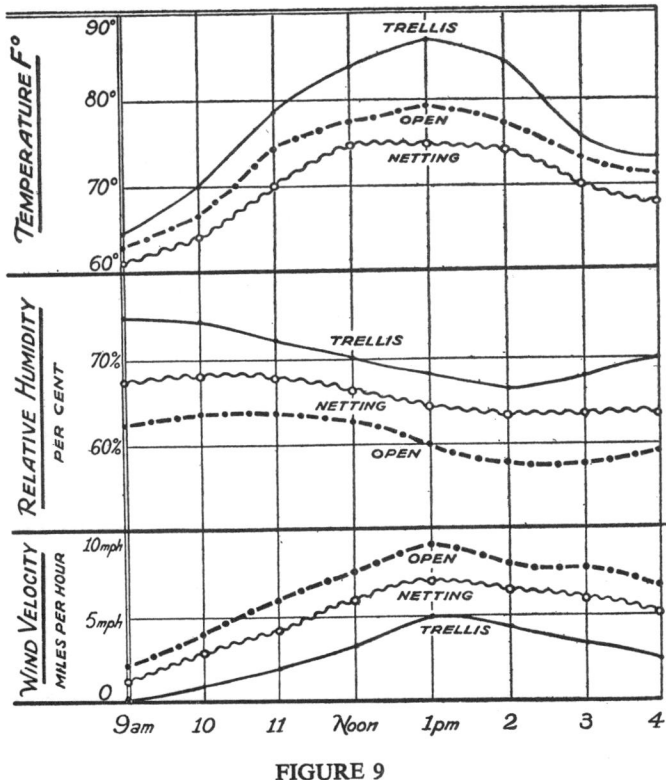

FIGURE 9

break than the wire netting and so had a considerably higher relative humidity inside it. The temperature inside the wire cage fell below that in the open and as much as 10° below that in the trellis cage. This appeared to be due partly to the fact that the wire acted as a sort of heat guard and partly to the reduced insolation owing to the small mesh of the wire.

These results agree well with similar work done by Stella S. Williams at the Grassland Research Station, Stratford on Avon in 1950. She experimented with a closed wire netting cage, and an iron hurdle enclosure fifteen feet square and three feet high, and found that the relative humidity and temperature inside the hurdle enclosure were both higher than outside. When the sun shone strongly the temperature inside the wire cage fell considerably below that of the air outside. Less dew formed inside the cage than outside, as the wire cut off some of the radiation from the ground beneath it at night. So it seems that in every satisfactory form of wind-break there is rather a nice adjustment to be made between not cutting off too much insolation, and yet having sufficient material to reduce the wind velocity enough to ensure raising the relative humidity. Very little protection from wind may be enough to produce surprising results. Timothy is normally a winter-green grass, but when sown on high ridge and furrow running east and west was only winter-green on the south slope of each ridge after a hard winter of frost and cold north-east winds.

It is the turbulent winds with gusts and lulls every five to ten seconds that cause so much structural and physiological damage to our plants. To steady winds the plants bend and receive no damage, but in turbulent winds the gusts are too sudden for the plant to conform to them and flowering heads are snapped off and stalks broken before the plant can give before the sudden increase of pressure. Physiologically steady winds may increase the water evaporation loss by about a 100 per cent, but a gust may suddenly reduce the temperature of a leaf over 20° below the temperature of the air. It is this direct action of the wind in removing water vapour from the leaf which so increases the water loss in comparison with still air, when a shell of damp air may envelop the leaf. If in addition the wind is cold the rate of loss of water is augmented, as the rate of supply to the plant is reduced by the chilling of the stems and root which suck up less water, and by increased viscosity of the whole plant which impedes the flow.

It is the break up of the eddies which are found in turbulent

WIND AND SHELTER

winds, and the reduction of the gusts they form, which is the main object of our wind-breaks; and as the eddies bounce over a solid wind-break but are broken up by one of open structure, it is a good thing to have a fairly solid wind-break to reduce the velocity of the wind, followed by an open one to break up the eddies.

A friend, whom I have recently interested in micro-climates, has just bought a new house and has a large garden on the south, east and west of it. To augment the shelter of the house and boundary walls, he is busy experimenting to see where he is going to put his various wind-breaks. Evergreen fences, rose pergolas, raspberry canes, and trellis are all to be employed as wind-breaks; so that when he has done there will be practically no wind, except from the south, that will be able to blow fiercely in the garden. I hope many more micro-climate enthusiasts will follow his excellent example; if so the number of Edinburgh gardens fit to sit in—in spite of our northern summer and constant wind—will multiply enormously.

Another interesting point is that the result of an experiment on wind-breaks becomes a sort of snowball; some crop, that we had not in mind when we began the experiment, is so pleased with its new environment that it flourishes exceedingly and in its turn shelters its neighbours, because masses of plants growing together provide a lot of mutual protection. In a few years the face of the garden may be completely altered, with a considerable increase in size and flowering capacity of the normal species. All the influences, of the rain, temperature, light, humidity and wind force alike, are altered, and finally we arrive at a garden in which there are a number of areas each with its own particular adopted plants, and its own micro-climate.

XII

Light and Shade

It is not pure chance that the predominating colour in nature is green. For if we grind up some cut grass in a solution of sodium chloride we obtain a solution of chlorophyll—a mixture of pigmented substances which make the grass green. And if we pass the rays of sunlight through a prism and produce a spectrum of all the colours of the rainbow, and interpose the solution of chlorophyll in the path of the coloured rays, we find that much of the long wave lengths at the red end, and of the short wave lengths at the blue end, of the spectrum are absorbed.

In the same way in nature the molecules of chlorophyll absorb the sun's light, and use the energy to combine hydrogen and oxygen, which the plant takes as water from the soil, with carbon taken from the carbon-dioxide in the air to form starches and sugars. These are first the chief sources of plant food, and are essential for the building of the cells of which the plant is made, and for the formation of the living content of those cells; after that they form the energy of all the animals that feed on green plants and eventually provide some of the energy of those of the human race who are meat eaters.

Absorption of the blue end of the spectrum produces chemical activity; absorption of the red end produces heat in the molecule. As heat encourages chemical activity, a molecule of chlorophyll is likely to show intense chemical activity. In fact a solution of chlorophyll fluoresces with a red light; the energy of the exciting sunlight is first used to produce heat and chemical activity, and the residue is re-emitted as light of a rather longer wave-length. The amount of energy absorbed by a green plant in making a pound of sugar is about 7,200 British Thermal Units,

LIGHT AND SHADE

and would raise forty pounds of water from freezing to boiling point. In an experiment in America, on an acre of maize for a hundred days from June 1st to September 8th, it was reckoned that about 20,000 pounds of sugar were made by the crop, requiring an energy of about 144 million British Thermal Units. The calculated energy of the sunlight absorbed for the same period amounted to 15,000 million British Thermal Units. So the green plant as a machine for converting solar energy into stored and usable energy is only about 1 per cent efficient, and yet from the point of view of food manufacture no more efficient method has so far been discovered.

If we take the incident energy of the sunlight as 100, then for an average laurel leaf the following figures give an idea of what happens to this energy.

Reflected from the leaf	15
Transmitted through the leaf	5
Used to make starch and sugar	1
Turned into heat and used to evaporate water from leaf	38
Turned into heat and radiated away to surrounding cooler bodies	25
Conducted to surrounding air in motion	16

By means of a thermo-couple it is possible to measure the temperature of a leaf accurately and compare it with the temperature of the surrounding air; it is also possible to get a good approximation to the actual temperature by wrapping the leaf round the flattened bulb of a specially sensitive thermometer. The variations in the temperature of a leaf are rapid as clouds pass over the sun in comparison with the changes of air temperature, and even on a cloudless day rapid changes take place as puffs of wind blow over the leaf. These changes are illustrated in the graph in Figure 10 where the values were taken from the leaf wrapped round the thermometer as described. By this method a maximum difference of 18° was found between leaf and air in comparison with 18° by thermo-couple, and 19° by theoretical calculation.

LIGHT AND SHADE

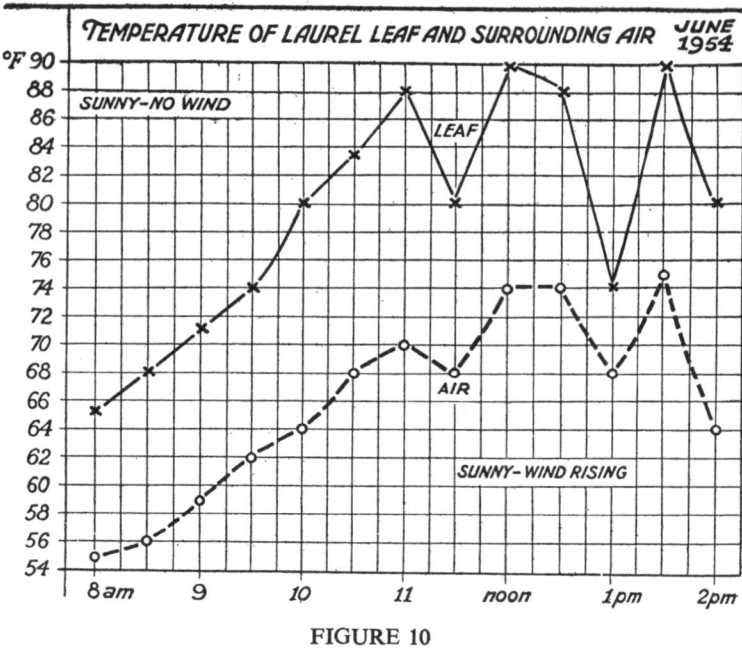

FIGURE 10

With a photo-electric cell it is possible to measure the reflected and transmitted light for a leaf. With a very young green laurel leaf the values are high at 20 per cent reflected and 10 per cent transmitted, but with older leaves the average is much lower at about 15 per cent reflected and 5 per cent transmitted. As so much of the sun's energy absorbed by the chlorophyll is turned into heat and dissipated by evaporation, radiation and conduction, and so little is actually used to make starch and sugar, it is not surprising that there seems to be no real lower limit of energy necessary for this process of sugar making, or photosynthesis as it is called. Some marine plants are able to carry on enough photosynthesis to maintain themselves at depths where the intensity of the light is only one-millionth of that of full sunlight at the surface of the water; so we need not think that any of the plants in our gardens and woods will suffer from lack of enough light to keep them green, though they may not get enough to produce flowers.

LIGHT AND SHADE

In March the light in oak and hazel woods is about one-eighth of the intensity several yards outside the edge of the wood; when the trees are getting leafy it is reduced to a twenty-fifth, and in June to about a fiftieth of the light outside. Close to the trunk of a single tree in full leaf it may only measure a hundredth of the light outside, while in the interior of a thick laurel bush the value may fall to a two-hundredth. Wood anemones and yellow archangel covered many acres of our wood in Northamptonshire and were an almost continuous flowerless carpet under the thick canopy of the trees, but in the clearings and where the wood had been coppiced recently they flowered profusely.

These values will not surprise the amateur photographer who knows only too well the great variations in the intensity of daylight even in the height of summer. On a normal sunny day with a cloudless sky about 80 per cent of the light comes from the sun and 20 per cent from the sky, so that the south face of the house getting both will be five times as bright as the north face which gets only the light from the sky. And as a white cloud is ten times as bright as the blue sky, it is possible for a day when white cumulus clouds are suitably illuminated by the sun to be brighter than a day when the sky is completely clear of clouds. This is particularly noticeable on the north side, where quite an appreciable rise in temperature above the average may take place when the sun shines on a number of white clouds.

As the temperature of the leaves of plants and trees rise above that of the surrounding air, the rate of transpiration of the leaves increases in consequence of the greater vapour pressure inside the leaf over the outside air, owing to the heating of the leaf. As laurel leaves may rise 18°, privet 14°, apple 18°, sunflower 20°, and orange 27° in full sun, the amount of water transpired is very considerable, and in the case of the sunflower amounts to above 700 units of water for every unit of dry matter produced. And throughout this intense activity the plant is all the time breathing in oxygen and giving out carbon dioxide and water-vapour as waste products of the process. In a bright light the plant takes in carbon dioxide from the air to form sugar five to ten times as fast as it gives it out in respiration; but since the plant also breathes at night, when sugar making is not going on, the rise in the stock of starch and sugar is a series of gains by day and losses by night with the nightly loss about half the gain by day.

These operations still go on in cut flowers, and as the water present in the stems is still drawn up after the flowers are cut air instead of water is sucked up, and when placed in vases the flowers wilt because their water-pipes are full of air bubbles. This can be avoided by hammering the ends of the stalks, or by putting the flowers in warm water first to displace the air bubbles. The falling of lupins and delphiniums, which after a day or two makes such a lot of work clearing up the fallen petals, can be avoided with freshly cut blooms, by putting a little starch in the water; with this the flowers will last a week, gradually losing their colour and dying but not falling even when shaken.

The visible solar spectrum is made up of lights of increasing wave-length from violet through blue, green, yellow to red. As light traverses our atmosphere the shorter wave-lengths are scattered in all directions by the molecules of the air, which on a dustless day gives the intense blue of the clear sky due to the scattering of the blue and violet. Just after sunset on a June night it is the light from this blue sky which floods our gardens and enhances the blues of the lupins and delphiniums in so enchanting a way. Just before sunset when the sun is shining through the maximum thickness of our atmosphere all the waves are scattered except the orange and red rays. The sun appears as a red ball and all the orange and red flowers glow brilliantly and the scents of many flowers and herbs reach their maximum fragrance. All plants make use of the sun's visible spectrum and in particular of these visible blue and red lights, but every gardener knows that some grow better in the sun and some in the shade.

The sunflower, a typical sun-loving plant and one which refuses to grow in the shade, appears to hold the record for sugar production with the maximum of about one-sixtieth of an ounce per square foot of leaf per hour, which probably worked out for that plant at about half an ounce of sugar a day. But the conditions for the test were not normal, as the plant was supplied with air containing 5 per cent of carbon dioxide, while the plants that grow out of doors have to make their sugar in air containing only ·03 per cent of carbon dioxide. At the other end of the scale the aspidistra, so fashionable as a house plant some years ago, only makes just enough sugar to supply its needs for growth. The difference between sun-loving and shade-loving plants must probably be sought in the effect of sunlight of considerable intensity on their production of chlorophyll and rate of production of sugar. High light intensity or high temperature and particularly both together destroy a certain amount of chlorophyll and so reduce the rate of production of sugar. Also owing to the small proportion of carbon dioxide in the air, this essential for sugar production is often the factor that limits the process.

We should therefore not be surprised to find that the leaves of the sun-loving plants are so constructed that carbon dioxide can diffuse into them very readily and quickly through their very numerous pores or stomata; sugar production is proportionately rapid, and the plant grows quickly and well, and throws up more and more leaves with fresh chlorophyll to replace the losses in the older leaves due to high temperature and light intensity. The leaves of shade-loving plants have much fewer stomata so that carbon dioxide does not diffuse into them so readily; sugar production is slower and could not afford to be still further slowed down by the loss of chlorophyll in bright sunlight. These plants would get no advantage by being in full sun, in fact a degree of shading that reduces the light intensity to about a third of its value in the open will enable these shade-loving plants to make sugar on a bright day as fast as their intake of carbon dioxide will allow. If we remember that Kansas State in the United States is known as Sunflower State from the prolific growth of the sunflower, there is an average temperature 10° higher than in London and a more intense sunshine, and that the aspidistra comes from the shady forests of the Himalayas and China, we see that they both are following their inherited characteristics, though growing in a land far away from their home.

LIGHT AND SHADE

Bright sunshine has curious effects on many plants. The goat's beard, cousin to the salsify whose thick tap roots we eat, is known as Jack-go-to-bed-at-noon, because its flower only opens about 10 a.m. and closes at noon. The flower of the evening primrose will not tolerate sunshine at all and only opens about sunset and dies before next morning's sun is up. Marrows on bright days throw up mainly male flowers and wait for the dull cloudier days to produce more female than male flowers. In my Derbyshire garden, ivy-leaved toad-flax grew all over a south-facing wall and a plant had managed to find a home in the mortar between almost every stone. I wondered how it managed to do this, as on a wall its habit of spreading underground was impossible. One day I took a close-up photograph of it with all its flowers turned towards the sun, a week or two later I took another showing how the ripe seed pods were now all turned away from the sun and towards the wall, so that when the pod split with explosive violence the seeds were thrown against the wall, and some found a resting-place amongst the cracks in mortar between the stones. But for this turning away from the sun when the seed was ripe all the seed must have fallen into the bed beneath the wall, and the plant would have died out on the wall itself.

As children we used a sunflower as a clock by fastening a cardboard collar, marked with the hours, on the stalk under its head, and a marked petal to act as a pointer. The flowering head turned towards the sun, keeping pace with its movement from sunrise to sunset with about half an hour's lag on sun time. This example of irregular growth which tends always to be greatest on the shaded side is common to many plants but is abnormally developed in the sunflower. At sunset the shaded side which has grown so rapidly during the day ceases to grow, and the unshaded side now has its innings, and in the dark humid air of the night grows rather faster; this turns the flower upright and rather more, so that it is ready to face the sun again in the east next morning.

Many of the plants in our garden come from all over the world; the hollyhock from China, chrysanthemum from Japan,

LIGHT AND SHADE

dahlia and cosmos from Mexico, phlox from Peru, and lilac and white jasmine from Persia, just to mention a few of the common ones. These in their native homes have very different lengths of day in summer than they have in Scotland where for about three months it is never dark at night. As plants are influenced not only by the intensity of light they receive, but also by the daily amount, we should expect that it would help them if we could arrange as far as possible for them to be given the number of hours' daylight they were accustomed to.

Most plants respond to the wave-lengths supplied by electric light quite readily, so that a number of very interesting experiments can be done in a small frame or greenhouse where electric light can be used to supplement the daylight, or used throughout the 24 hours after cutting out the daylight. Lettuces grow well when given 6 to 12 hours out of the 24, but bolt and form seed heads with 18 to 24 hours' light. Tomatoes fruited with 6 to 18 hours' light but died when exposed to the full 24 hours. Geraniums flowered when given up to 12 hours' light, with more than that they faded off and gradually died. The sunflower, as we might expect from what has already been written about it, delighted in 24 hours' continual daylight.

Love-in-the-mist is a typical plant which is greedy of daylight; winter-sown in electric light it will flower weeks earlier than when sown in the spring, and when planted out will come into bloom during the period when it is light all night. If it is forgotten and only sown late in spring out of doors it may not come into flower at all as the day will be only about twelve hours long by the time it is mature; the vegetative growth will be enormous but we shall have no flowers. Cosmos, by contrast, likes long nights and will not flower well during our summer months in Edinburgh when it is light all night. Planted in the open, it tends to defer flowering until the shorter days of autumn, and then may be cut down by frost before it has flowered at all. If we grow it where we can artificially limit its day to twelve hours it will come into bloom quite readily early in the summer. Other plants that benefit by the same treatment and can be made to flower early by artificially shortened days are chrysanthemums, dahlias, and sweet williams. But it is important to note that when daylight has to be limited to twelve hours the other twelve must be completely uninterrupted darkness; a few minutes' light will undo all the good done by the hours of darkness.

When using artificial light to control flowering of short-day plants, and to induce flowering of long-day plants, the intensity of illumination necessary is very low in comparison with full sunlight. A 100-watt lamp giving about one candle power per watt will provide about ten foot candles over quite a sizable bench three feet beneath it. This lighting is very economical and costs us about a penny for a twelve-hour night.

XIII

Cloches and Frames

As we have already seen, the radiation that the sun pours on the earth is partly used in evaporating moisture from the vegetation and the ground, partly in heating everything on which it falls, and the residue is re-radiated on a much longer wave-length than the incident sunlight. It is the peculiarity of glass that it passes 90 per cent of the incoming sunlight, but hardly any of the outgoing long wave re-radiation, that makes it so valuable for greenhouses, frames and cloches, and which constitutes what is known as its 'greenhouse effect'. Inside the glass enclosure the temperature mounts up and so little of it escapes that the air temperature over an area covered with greenhouses is lower than the air temperature over the open ground all round.

Sixty years ago I saw my first cloches in my grandmother's garden. They were of the old French pattern, made of thick glass in the shape of a bell, with a knob on the top to carry them by through which a hole allowed a little ventilation. I was so interested in the cloches themselves that it took some time before I realized they had a purpose, and did full justice to the few early strawberries I found beneath them. They were obviously clumsy, heavy to move, cut off much of the sunlight owing to the thickness of the glass, and were prone to much condensation owing to poor ventilation. Back at home I hunted round for something to use as a cloche but could not find anything suitable; so I got a large biscuit tin and cut out the bottom and windows in the sides, over which I fixed glass in metal grooves made by folding over a piece of the metal at top and bottom of the windows; with a sheet of glass in place of a lid I had a tolerable box cloche under which I grew seeds and ripened early strawberries.

CLOCHES AND FRAMES

But in due course the novelty wore off and it was only many years later, when I was doing some work at Chertsey, that I realized what could be done when I saw the results of Mr. J. L. H. Chase's researches into making modern cloches of all kinds. Mr. Chase has described the evolution of these modern cloches under the hands of his father and himself in his book *Cloche Gardening*, and it was thanks to his kindly co-operation that I have been able to make some experiments with cloches during the last few months.

Gardeners have always wanted some protection against the worst handicaps of the weather—too strong a wind, too hard a frost, too heavy rain—and the damage done by birds, cats, dogs and rabbits. In addition to all these advantages they wanted something that would give free passage to light, be easy to move, yet keep their shape and be secure in a high wind, allow good ventilation, and not require moving when watering was necessary. It gives a good idea of the importance of the invention of glass cloches when we can say that they provide all these requirements.

In our Scottish climate the north-east wind blows strongly for months in the spring, and wind shelter is one of the essentials for successful gardening; and it is a great advantage to be able to couple this with an increase of temperature of up to 8° in early spring under a glass cloche in dull windy weather. Protection from frost is even more needed than in England, for up here snow in May is by no means rare and damaging frosts last much longer than in the south, and sometimes until early June. Cloches do not make heat, they only trap the sunlight and store it up against the chill of the coming night. So in winter when the sun is low in the sky or does not shine at all, the temperature under a cloche is only a few degrees above that in the open air; but as the sun rises in the sky and becomes more powerful the difference between cloche and air temperature rises rapidly, and the minimum under the cloche on the morning after a sunny day is very dependent on the maximum temperature that day. For instance, if the maximum cloche temperature was 21° above the maximum air temperature, then the minimum cloche temperature after a calm clear night would be just under half 21°, or say 10° above the open air minimum.

My experiments, begun in November 1953, on the differences of mean air temperatures inside and outside a cloche, show clearly that the difference increases with the increase of altitude of the sun and the number of hours it shines per day. Winter for the cloche gardener is limited to the months of December and January, every other month of the year having a mean temperature under the cloche of 42° or over, so that growth begins in February and ends in November. The mean air temperature under the cloche is always higher than that of the open air and ranges from about 3° higher in December and January to 15° higher in May. And these figures can be improved upon if the gardener has the chance to put his cloches at the foot of a south-facing wall with a projecting canopy fixed above them. Then the mean air temperature under the cloches will be above 42° all through the winter, and will range from about 5° higher than the mean open air temperature in December and January to 20° higher in May. Sometimes the gain in warmth under a cloche may be quite spectacular, a glass cloche under a south-facing wall on 19th May 1954 achieved a range from 52° to 114°, with a mean temperature of 83°, which was 30° higher than the mean air temperature in the open.

From January to May 1954 inclusive there were sixty frosts in the open air and only fourteen under the glass cloche, of which twelve took place in January and February. So that after spring had begun the glass cloche was practically frost proof although the open air temperature fell to as low as 23°. On the few occasions when frost did penetrate into the cloches no damage was done, probably because the plants were dry and so could stand up to frost much better than when they were wet. Also the ground surface was dry and formed a badly conducting layer so that the frost did not penetrate into the soil under the cloches. Outside where the soil was beaten down by treading and rain the frost penetrated several times during the winter to a depth of four inches.

It is a great satisfaction to know that one's plants are safe from birds and beasts under cloches. Though we have no cat or dog, those of our neighbours always seem to choose our garden to scratch in; the rabbit which lives each winter under my compost heap has to feed himself, and sparrows every spring ruin most of my polyanthus, but pay back handsomely in the greenfly, caterpillar, and daddy-long-legs season later on. This year a young rat bit off every flowering head on several roots of pyrethrum before I discovered he also had made a home for himself under the compost heap. So I cloched the rest of the sprouting pyrethrum, and made a scorched area where there was nothing to eat round his hole except some tempting baits of bread and butter and Rodine. He fell for them, and I picked up his dead body next day.

Though lateral movement of moisture in ordinary soil is slow, the same movement in manure or compost is much more rapid. If the ground under the cloches is manured or composted over an area greater than the width of the row of cloches the rain running off the glass seeps into this manure layer and travels laterally under the cloche from both sides. Even when the soil surface under the cloche appears dry there is plenty of moisture a few inches down, and it is never necessary to remove the cloches and water and so destroy the tilth of the soil. If water is needed during a drought it can be poured on to the glass without moving the cloche at all, and if the ground between rows of cloches is kept hoed it will absorb all the rain that falls and prevent losses by run-off or evaporation.

When plastics were invented it was natural to expect that they might replace glass for cloches owing to their lightness and apparent immunity from breakages. But in practice it was found that plastic had more disadvantages than advantages. To be reasonably cheap and able to compete with glass the plastic cloches had to be made of very thin material and so were unable to keep a rigid shape; they very easily scratched and weathered, and were too light to be kept in position without tethering which was a great nuisance when the time came to move them. Even those with legs to be pushed down into the soil in place of tethering were often blown all round the garden in a strong wind. Also they were habitually covered with condensation and this, as will be seen later, seriously reduced the amount of light they transmitted. With the larger glass cloches one sheet of glass could be removed to give access to the crop for picking or pollination; the plastic cloches were not rigid enough to allow this and the whole cloche had to be removed to get at the crop. Ventilation could only be achieved by leaving an open space between each cloche and this was risky in a high wind; each glass cloche has its own ventilation at the ridge and eaves which gives sufficient ventilation over most of the year without spacing apart in the rows. In gardens where young children play plastic cloches are safer, but many advances in their design will be necessary before they can compete on equal terms with the older glass cloches.

Ordinary horticultural glass and unweathered plastic each absorb about one-tenth of the incident light. The visible part of the sun's radiation represents rather more than half the total energy, and ultra-violet and infra-red rays that we cannot see make up the rest. Ordinary glass transmits all these with the exception of some of the ultra-violet rays which are not used by plants, and some of the infra-red which it only transmits feebly. Unweathered plastic passes the ultra-violet rays which glass absorbs and the infra-red rays which glass transmits feebly, but over the whole range of the visible rays plastic does not transmit the sunlight quite as well as glass. Beyond the infra-red rays of the sun lie the long infra-red rays re-radiated from the earth; these are completely trapped by glass, but not nearly so completely by plastic, so glass has a superior 'greenhouse effect' to plastic and the temperature under glass cloches is always higher at night than under plastic, the difference in favour of the glass amounting to about 2° to 4°. Also due to the loss of long infra-red waves through plastic the maximum temperature in the daytime under glass was always about 3° to 4° higher than the maximum under plastic.

Except in times of drought, plastic is almost always covered with condensation; this reflects some of the incident sunlight and reduces the day maximum by another 3° to 4°, so that the

CLOCHES AND FRAMES

maximum under plastic is often 6° lower than under glass. All this is borne out by the light intensity measurements over five weeks in the spring of 1954; a new plastic cloche after being

TABLE 11

Light and Temperatures Under Glass and Plastic Cloches and in Open Air

Date 1954	Glass		Plastic		Open		Light Values			Weather	Remarks
	min	max	min	max	min	max	Open	Glass	Plastic		
	deg.	deg.	deg.	deg.	deg.	deg.					
Mar. 24	40	52	38	50	34	44	375	335	300	Sunny Intervals	
25	36	56	35	54	33	50	30	28	24	Dull cloudy all day	
26	38	65	36	63	36	52	400	350	300	Sunny Intervals	
27	36	68	34	66	32	54	400	360	325	Sunny Intervals	Plastic
28	40	67	38	62	36	52	250	225	175	Sunny Intervals Rain	Cloche covered
29	35	73	33	70	30	59	100	100	75	Cloudy, Rain	with condensation
30	42	74	40	68	38	52	200	175	150	Sunny Intervals Rain	every day
31	34	73	32	70	28	52	300	275	250	Fine Sunny	
Apr. 1	34	78	32	76	28	54	325	300	250	Fine Sunny	
2	42	64	40	62	35	52	50	45	40	Rain then cloudy	
3	42	61	41	56	38	52	32	29	24	Rain then cloudy	
4	38	76	37	70	35	50	400	360	300	Fine Sunny	
5	36	74	32	72	28	54	300	275	250	Fine Sunny	
6	36	82	33	77	30	58	400	360	300	Fine Sunny	
7	37	77	34	74	30	50	400	360	360	Fine Sunny	
8	43	86	41	80	39	66	450	400	400	Fine Sunny	Plastic
9	40	78	38	73	34	63	400	360	360	Fine Sunny	Cloche
10	34	74	32	70	29	60	225	200	200	Sunny Intrvls	demisted
11	46	83	44	79	42	64	500	450	450	Fine Sunny	
12	41	80	38	76	36	58	500	450	450	Fine Sunny	
13	42	82	39	78	36	64	300	250	250	Fine Sunny	
14	45	84	44	78	42	66	300	275	250	Fine Sunny	
15	47	92	45	86	42	70	600	550	500	Fine Sunny	Plastic
16	37	91	34	86	31	70	150	150	125	Cloudy, Rain shower	Cloche becoming
17	47	61	46	59	42	52	33	30	25	Fog and Rain shower	covered with condensation
18	42	83	40	79	38	68	500	450	350	Dull a.m. Sunny p.m.	again

rubbed over with a demisting compound passed as much light as glass, but when covered with condensation was much less efficient, and while glass nearly always passed nine-tenths of the incident light plastic has often only passed seven-tenths. When it has become scratched and weathered, and is tending to get out of shape and fall into slight folds and hollows, it may only pass about half the incident light. Demisting on a commercial scale is of course quite impossible, and in any case only lasts a couple of days, so this continual condensation is a heavy drawback to plastic cloches. Cloches can be moved so easily that we can change their position in the garden several times a year, so that the soil under them does not become exhausted by continual use. Most frames have to be renovated with fresh soil at frequent intervals, but I have a small portable frame which I can take down and put up again a few feet away in about a quarter of an hour and so allow the weather to renew my soil for me. But, in comparison with cloches, frames get very little sunlight in winter, because the sun is low in the sky and throws long shadows of the sides of the frame; and as the light only comes into a frame from above the plants tend to grow spindly if the weather is poor and the glass cannot be removed for a time every day. So frames need constant ventilation and watering, and when a crop has been sown it has to stay in the frame till harvested or transplanted, which means that the frame is occupied with one crop for a long time.

In spite of all these disadvantages a small frame is most useful, especially if a soil warming apparatus run off the mains is incorporated in it. Heating on a large scale in a greenhouse fourteen feet by ten feet with a span roof is very expensive, and to keep such a greenhouse up to 50° all the year round costs about £20 for electricity at a penny a unit, but we can keep a frame about 15° above the temperature of the outside air throughout the winter for about a shilling a week after an initial outlay of £5. The frame may be used as a hotbed, with the heating wires warming up the whole mass of soil to grow early lettuces and carrots at a tenth of the price charged in the shops, or as a propagating bed where seeds of garden annuals and tomatoes can be raised, dahlias sprouted, and cuttings made from chrysanthemums.

Heating the soil at a depth of a few inches to warm the surface layers is of course exactly the reverse process to what happens every day when the sun shines and the surface warms up to heat the lower layers. The hot surface gives away only a little heat to the badly conducting air above it, and we want to copy this and put a badly conducting layer below our heating element so as to throw most of the heat up rather than down. For this peat or grass turves root side up make a poorly conducting layer on which we can lay our heating element. This element will probably be about forty-five feet long and has to be spaced apart about four or six inches and to get this done neatly in the enclosed space of a frame is a formidable business as the wire tends to kink and tie itself in knots. I found that a framework of bamboo canes tied together and laid on the lawn where there was plenty of elbow room was a great help. The wire could then be attached to the framework at the required distances apart with insulating tape, and when the job was complete the framework could be lifted and laid in the frame and the connecting plug inserted into the mains socket. The wire is then covered with ordinary garden soil, adjusting the depth of the soil to suit our requirements. About six inches of soil for a hotbed, and about two or three inches for a propagating bed, are suitable depths, as we need the wire fairly near the surface when we are going to put boxes of seeds on the top of the soil.

When used as a hotbed the soil will need renewing every autumn and a fresh layer of compost and soil put into the frame. As a propagating bed, where our seeds and plants are in boxes and plots, there is no need to renew the soil so long as we keep it damp enough to make for easy conduction of the heat from the element to the surface. But in either case the garden is no place for the amateur electrician; and a soil heater may be a source of danger—especially in a hotbed where digging and cultivation with metal tools is a necessity—unless the whole installation is set up by a competent electrician who will see that it is properly insulated and made waterproof and shockproof. This is all the more important because the older method of soil heating, by means of a step down transformer which reduced the mains

CLOCHES AND FRAMES

voltage to about twenty volts which was quite harmless, is now replaced by the more efficient full mains voltage heating. The heating wire is well insulated and covered with plastic which is not affected by acid, alkalis, water or mineral fertilizers and is not attacked by insects or fungus. But of course it can be cut with a sharp tool, and at full mains voltage can be dangerous if not properly fixed up.

The ordinary cold frame is dependent on the amount of sunshine it gets and cannot be controlled by a switch like the heated frame, but its rise in temperature above the outside air temperature is by no means to be despised. The estimated gain in earliness in the frame remained fairly constant at about three weeks all winter and spring. The rise in the maximum temperature in a south-facing frame was largely due to the amount of sunshine between 11 a.m. and 1 p.m., when the sun gets into the frame without throwing long shadows of the sides. If we get these essential hours sunny we can probably achieve an increase of maximum temperature above the outside air of 10° in February, 15° in March, 20° in April, and 30° in May and June. If we have achieved a good increase in maximum temperature during the day, we need not fear a frost at night, as the minimum air temperature in the frame will remain above the outside air minimum by about one-third of the increase in the maximum. So that after a sunny day our frame minimum will be about 3° in February, 5° in March, 7° in April, and 10° in May above the outside air minimum. And if we care to cover our frame with sacking we can safely add another 2° to these figures. During the winter of 1953–4 we had sixty-five frosts in the open air but only seventeen in the frame, and with a sacking cover on all frosty nights the lowest temperature was 26° when the outside temperature was 14°.

The most dangerous nights are the calm clear nights that follow a day of rain or no sun, and this is where the heated frame scores. Because we can, by the turn of a switch, heat up our soil to take the place of the sun for a few hours before bedtime, and then switch off safe in the knowledge that for several hours this heat will be rising to the surface to ward off the frost.

XIV

Micro-Climates on the Farm

In our gardens the plants have to compete with each other and the weeds and make the best they can of the soil and the weather, and in doing so have considerable help from the gardener. On our farms the crops have just the same things to contend with and can be given little help by the farmer, and the grass crop has to compete with the grazing animals as well.

On our Northamptonshire arable land the continuous growing of corn crops promoted weed infestation with couch grass, charlock, poppy, corn marigold, buttercup and bindweed. These could be eradicated by the intensive cultivation for the root crop in the rotation, and by the hoeing and singling of that crop, but a succession of root crops favoured the growth of fat hen, groundsel and fumitory. One of the chief reasons for a rotation of crops is the eradication of farm weeds.

Weeds in pastures were a more difficult problem because here the preferences and different methods of grazing of the various animals came into play, and fields grazed entirely by bullocks became infested with buttercup, ragwort and ox-eye daisy, but when sheep were put in with the bullocks all these weeds soon disappeared. Excessive treading always produced large crops of daisy, plantain and silver weed, and these grew in every gateway and on every field grazed by an excessive number of stock or not given periodic rests. Undergrazing was no remedy, for that favoured thistles, ragwort, and yarrow, which had to be pulled out or cut with a scythe or mower before they seeded.

Rabbits destroyed the good grasses—especially rye-grass and clover—and these were replaced by moss and weeds. A pasture seriously infested with rabbits soon lost about half its value, as was shown by the reduction in live weight gain of the animals pastured on it. Geese, in contrast, by their close grazing and lavish manuring produced a level and lawn-like sward. Any mistake in the management of a pasture was at once made evident by the lack of relish of the grazing animal, because, by the change of environment, the pasture had lost some of its succulence and palatability. These qualities, like the 'good heart' of the arable land, defy exact definition, but are of great importance to the successful grazier. For any enthusiast in search of new work, the study of the micro-climates of a grass sward, and the succulence and palatability of several grasses, clovers and herbs growing together in a pasture, is certainly a rewarding piece of research.

Prehistoric man knew of the seasonal changes of succulence and palatability of pasture plants. He realized that every grass, herb or shrub was palatable at some time of the year, and on his shielings, bilberry, gorse, heather, cotton grass and heath rush were sought after in their season, and the change of pasture put a bloom on the coats of his animals that would never have been gained on the pastures of the homestead.

Of all farm animals cattle are the most fastidious grazers, but different breeds behave differently and individuals in any one breed have tastes that differ widely from the rest. This has been borne out in this country where Sussex cattle often eat all before them without much discrimination between grasses, herbs and weeds, while Devons pick and choose and often wander over the whole pasture before settling down to graze when put in a new field. Cows and bullocks appear to graze differently and heavy milkers are often more fastidious than their less productive sisters. To a few lazy individuals accessibility counts for more than palatability.

The anonymous author of a book on lucerne, sainfoin and burnet, published in 1775, had much to say on burnet. It was an evergreen which resisted cold, heat and drought better than any other forage plant and was ready for use at all times of the year and was particularly useful in March, April and May when winter foods were exhausted and pastures not yet ready. If not grazed too late in the autumn it made excellent foggage, cows and sheep were very fond of it and did well on it.

Arthur Young first cultivated chicory in England in 1788 when he sowed it along with clover, trefoil, ribwort plantain and burnet, under barley. He also sowed it in drills a foot apart and it produced thirty tons of green food per acre. It was so liked by fattening bullocks, cows and sheep that it was always grazed close to the ground; in spite of this it was very persistent and lasted ten years. Its root went down thirty inches and took its food from a greater depth than the grasses and clovers which flourished right up to the stems of the chicory.

After this little was added to our knowledge of the palatability of forage plants until 1885 when Dr. David Wilson—who later became Sir David Wilson of Carbeth—worked on the chemical composition of grasses and herbs at various stages of their growth. All previous work on this subject had ignored the seasonal effect on succulence and palatability and Wilson put the whole thing on a scientific basis in the words, 'I have observed that the taste of animals for the different grasses is in accordance with the results of analysis if these are fairly interpreted, and that animals tend to graze each grass when it is at its most succulent stage of growth.'

My father, who was a pioneer in most details of agriculture, was sceptical of this too tidy clearing up of all the difficulties which surrounded the likes and dislikes of his bullocks, and he organized an experiment on the behaviour of his animals. He used three fields adjoining each other and sown down to leys; the first was a new cocksfoot, timothy, meadow fescue and clover mixture with a herb strip of ribwort plantain, chicory, burnet and clover on the headland. The second was a two-year-old rye-grass and clover ley with no herb strip. The third was three years old, of the same mixture as the first, but with no herb strip and which had been shut up for foggage after hay harvest.

The gates of the fields were opened and forty bullocks turned in to graze the three fields as they liked in the spring of 1894. The result was somewhat surprising, for thirty-eight out of the forty bullocks at once started grazing the foggage and stayed there for three days and nights. When they were driven out and the gate was shut they collected on the herb strip of the first field while the remaining two bullocks grazed the rye-grass and clover ley of the second field. These two were shy animals which had always preferred to graze by themselves. Nothing would persuade the thirty-eight bullocks to graze the rye-grass and clover ley except actually shutting them in the field, and even then they spent a good deal of their time trying to get through the hedges to the foggage and herb strip. I remember Father's comment was short and to the point, 'You can't decide what bullocks like in a laboratory; this will upset Dr. Wilson, I fear'.

Recently those who had provided their cattle with luscious leys of the new Aberystwyth strains of grasses and clovers were sometimes surprised to find that their animals deserted these leys to seek out herbs and rough grass from the hedgerows or from the herb strips which many farmers grew on their headlands as a variation from the plain rye-grass and clover sward of the rest of the field. By 1948 several investigators were trying once again to devise experiments to show what grasses or combinations of grasses and herbs were most favoured by bullocks and cows. These trials do appear to show that ribwort plantain and timothy are unusually palatable to numbers of cattle, cows and sheep during a great part of the grazing season and not only at the season of their young growth, and these preferences are reflected in the live weight gains made. The evidence for the palatability of ribwort plantain is very striking. During one test it seemed as if it was first and all the rest nowhere. Even when flowering heads had appeared the animals consumed the stems as soon as the leaves were eaten. It proved especially resistant to drought and while the majority of the grasses were suffering it still threw up green and succulent herbage; even in a wet season it had a very high measure of inherent palatability which still kept it as firm first preference.

Though some animals, fed on too luscious and succulent growth in the spring, do seek out roughage along the hedgerows and so avoid bloat, and some will use mineral licks provided when the herbage is deficient in minerals, yet we can hardly endow cattle with such an acute sense of the chemical make-up of different pasture plants that they can detect with certainty the difference between two grasses having only minor differences in their chemical composition. It seems as if something much simpler, some acute sense of touch and smell perhaps, may guide the animal to those grasses or herbs for which it shows a decided preference.

On the principle that 'a little bit of something we like is good for us' it may be that these preferences make the difference between one that fattens quickly and an average beast. For the animal's rumen is not only a chemical laboratory, and the process of chewing the cud may be infinitely more pleasurable and satisfying when it means chewing something which had tickled the animal's palate in the first instance.

But in spite of the advances in our knowledge of the chemical composition of foggage plants, and the efforts of our farmers to produce succulent leys, herb strips, foggage for winter feed as well as hay, dried grass and silage, we still have to admit that we know little more about the likes and dislikes of our animals than our ancestors did 2,000 years ago when they sent their cattle to the summer pastures on the shielings to put a bloom on the coats of their bullocks and more milk in the udders of their cows.

A study of a row of seedlings, ready to be thinned, in our garden will throw a lot of light on the farmer's methods of producing a new ley. If we examine the row critically we shall notice that it probably consists of a few well-grown seedlings down the middle of the row, flanked on either side by slightly smaller ones, and a number of very small ones on either edge of the row. The weak spot of a seedling is where the stalk issues from the ground; and, if we examine the curves for bare soil in Figure 8 of Chapter IX, we shall see that the small seedlings at the edges of the row probably had to endure a temperature of 98° at that weak spot on the day the graph was made, while the seedlings in the middle of the row only had to compete with a temperature of about 70°.

If now we keep the best-grown seedlings in the middle of the row and pull out all the rest, then those that are left are suddenly bereft of all the shelter which enabled them to outgrow their neighbours and will have to face a high temperature at their weak spot. This will certainly upset their growth and may well prove fatal. But on a small scale in our garden we can either water the row of seedlings regularly so as to give plenty of moisture to be evaporated and reduce the temperature of the surface of the soil, or we can provide some temporary shade until such time as the seedlings have grown large enough to give shade to themselves and each other.

But what is the farmer to do when his fields of roots are ready to be singled; he cannot water or give shade, and if he singles when the ground is dry and the sun is hot he runs the risk of big losses by 'strangles', as he calls the result when the singled plant falls over and dies, with every appearance of having been strangled just at ground level. At home we singled roots in the cooler weather when the ground was damp and there was no hay fit for stacking, we never singled them in hot dry weather which was more suitable for haymaking. But other farmers, who did not believe in my experiments or that strangles had anything to do with the sun, suffered heavy losses when they singled their roots when the ground was dry and the sun hot. They rightly claimed that the hot sun killed all the rubbish but if it also killed the roots this seemed a doubtful advantage.

On our heavy soil it was impossible to keep bullocks on the pastures in the winter, even if the weather was mild enough, because of the damage they did to the turf, so during our experiments with rows of seedlings it occurred to us to grow some grass in rows far enough apart to encourage the bullocks to walk between the rows when they grazed them. Farm horses all learnt to walk between the rows of crops, so why should not bullocks learn too. This would prevent damage to the grass and at the same time kill the weeds between the rows and save cultivation.

One spring we sowed a small field with a fairly heavy seeding of timothy in rows two feet apart, and left it to grow all summer and autumn until it became the foggage which was just then

MICRO-CLIMATES ON THE FARM

rather the fashion, but with this difference that we hoped our foggage would all be eaten and not trodden into the mud. By December, after a frost or two, there was a lot of fodder which the cattle appeared to relish, and to our satisfaction hardly any of it was wasted, as the cattle from the beginning kept between the rows when grazing. Whether it was an economic proposition or not we never solved, for shortly after, on doctor's orders, Father gave up arable farming and only kept on the pastures, and so we had no opportunity of going on with the experiment which is now being repeated with different grasses at the new grassland research stations.

In our droughty East Midland climate in summer it was often difficult to get a good take of grass and clover seeds when sowing down a ley. We found when we sowed seeds in rows that unless they came up fairly thickly they did not get enough protection from the sun, and the ground soon got too dry for growth; this happened especially quickly with timothy and clover, as this mixture seemed particularly sensitive to adverse weather conditions. So we aimed at getting about 200 seedlings per square foot in the early stages when protection was really necessary if a heat wave of over a week or so came along before the seedlings began to tiller. After that the plants were more bushy and provided shade for their own roots so that they could stand any reasonable drought without many fatalities, though growth was considerably checked.

To avoid all these difficulties farmers in our district mostly sowed their grass and clover seeds under a crop of short-strawed oats. This so-called nurse crop came in for much adverse criticism by farmers in better districts which seldom had a drought, and they called it a robber crop instead of a nurse crop on the plea that the oats took all the moisture and left the grass and clover worse off than without them. But this argument was fallacious, because the oats' roots took their water from a layer far below the grass roots and so did not compete with the grass at all. And they were literally a nurse crop, as by their extra height they shaded the grass and clover seedlings and lifted the zone of extremes of heat and cold well above the tender young

MICRO-CLIMATES ON THE FARM

plants. The thickness of the oat crop also produced extra humidity at ground level and acted as a wind-break to check evaporation and so combated the drought. Nurses are needed only so long as their charges are young, and the oat crop could be dispensed with after a month or two and could be cut green for fodder as soon as the danger to the grass and clover was over.

To get 200 timothy seedlings per square foot without a nurse crop of oats entailed using about 35 lb. of timothy seeds per acre. It is extraordinary how little cover to the ground this number of seedlings gives, only a little over one green spike on every square inch, so that for a start the ground is hardly shaded at all. Yet temperature readings show that even this sparse cover reduces the temperature near the soil surface a little below that over completely bare soil, and this reduction in temperature steadily increases as the seedlings grow. By the time the seedlings cover about two-thirds of the ground the field becomes a patchwork of different temperatures, high where the cover is still sparse and relatively low where the cover is good. When this stage has been reached the critical time of growth is over and the crop should be secure in normal years.

This description of the establishment of a new ley is based on results in Northamptonshire only; I have not done such microclimate experiments elsewhere on different soils and in different climates. But more experiments on these lines to decide the minimum amount of seed per acre that can be used with and without a nurse crop, and which will still provide good establishment after a drought are well worth doing by anyone who has the patience and opportunity to do them.

Fifty years ago it was the common practice amongst farmers who had the famous Leicestershire fattening pastures, around the Welland valley near Market Harborough, to collect the bullocks' droppings every week into a heap in the corner of the field, and at the end of the season when the bullocks were taken off the pasture to re-scatter them over the whole field. This practice was costly, but prevented the growth of coarse patches of grass which were not properly grazed, and enabled the farmers to attain that billiard table level of grazing all over their fields which was one of the marks of these wonderful pastures. But in addition to this it was the firm belief of these farmers that, quite apart from the manurial value of this top dressing, spring growth of grass started much earlier in the manured fields than elsewhere. And as a part of the recent centenary celebrations of these Midland fattening pastures I was asked to find out if there was any scientific backing for this popular belief.

I first arranged with a friend that when he spread his droppings over his fattening pasture in the autumn he should leave a good-sized patch without any of the droppings, and give it what he considered an equivalent dressing of artificial manure. In this way, as far as possible, we eliminated the manurial effect as opposed to the increased warmth effect from a dressing of dark coloured droppings. Then I set to work on my own lawn to find what extra warmth was produced at a depth of four inches by a dressing of farmyard manure. This was a very troublesome experiment, as during the late summer and early autumn the grass grew fast and the lawn had to be brushed and mown and the manure renewed several times. But the results showed that we were on the right lines as the manure evidently increased considerably the range of temperature at the 4-inch depth.

So I packed up two pairs of my wrynecked maximum and minimum thermometers, and asked my friend to dig a pair into each part of the pasture, one pair under the spread droppings and one pair under a clean patch. We started taking readings at the middle of March and went on to the middle of April, and the result was a conclusive verdict in favour of the farmers' belief that the manure encouraged earliness of growth apart from its manurial value. Out of thirty days, during which there were fifteen frosts on the part with no droppings and only seven on the part with them, the mean temperature at the 4-inch depth beneath the droppings was only below a growing temperature of 42° on five days, while on the other portion it was below 42° on seventeen days. Also the difference in growth of the grass was so marked that by the middle of April it looked as if the plot that had only had artificial manure had been mown, in comparison with the rest of the field.

XV

Micro-Climates in the Home

When some of my thermometers were not in use in the garden it seemed natural to employ them indoors to see how the temperature and humidity of some of the rooms of my house depended on the outside conditions. There are so few records of micro-climates in the home, on which so largely depend the cost of heating it in the winter, that there is a great need for this type of research.

My house is a bungalow of forty by fifty feet, facing approximately north and south, so that two rooms get sun most of the day while the two rooms at the other end get only a little sun in the early morning. Bungalows, with their large expanse of roof in comparison with their size, present special problems. It is well worth while having the hot and cold cisterns and pipes in the roof lagged with asbestos or felt as this gives reasonable security against frost and ensures that the hot water of the night before is still warm enough to be useful the following morning. If the expense can be afforded the whole floor area of the loft should be covered with some non-conducting material so that the warmth of the house cannot escape easily into the roof; this will make it much easier to keep the house warm in the winter, and a neighbour, who has had it done, tells me that he saves over £4 a year on the cost of keeping his house comfortably warm.

Many people do not realize that heating systems are not used to warm us but merely to ensure that we lose the heat of our body at a comfortable rate. The temperature of the human body is kept up by the food we eat and our thermostatic control tries to keep it at 98·4°. The food also supplies all our energy, both mental and physical; mental activities use up comparatively little energy, hard physical work uses up a great deal. But in the process our body generates heat, which has to be dissipated by radiation, conduction, and evaporation; and if we are to be comfortable we must lose heat at the same rate as we generate it. A lightly clad sedentary worker in a room where the air is relatively moist and free from draughts is comfortable when the temperature is between 60° and 65°; but if a window is opened letting in a draught, even though the outdoor temperature is the same, or an electric fan is turned on, he would feel chilly at once because his rate of evaporation and conduction to the air is increased by the air movement. He is also radiating heat from his body to all the surrounding objects such as walls, ceiling and floor when they are at a lower temperature than he is. In our climate this means that he radiates heat summer and winter, though much more rapidly in the winter when everything in the room is so much colder.

So the problem of comfort in our home depends on how much heat our rooms need from the sun and air in the summer, and from our fire or heating system in the winter, so as to keep the temperature of the walls, ceiling and floor, and the air of our room as near 60° to 65° as possible. In the summer we have to prevent the heat of the sun getting into our rooms, in the winter we have to conserve the heat produced in our room against loss to the external air. Both of these we do to some extent by having walls at least nine inches thick and up to twelve inches or more if our house stands high in a windy locality. The absorption of summer sunshine by our outside walls is reduced if they are colour-washed in yellow, and light-coloured inside walls pass heat to the outside less readily than dark or wood-panelled walls. Outside walls should shed rain and dry quickly, because wet walls in winter cool by evaporation and pass heat about ten times as quickly as dry walls. The material of which walls are made is important, and as air is a bad conductor we find that dry porous materials containing a lot of trapped air are much better than close-grained stone, cement, asbestos or metal. Nine inches to a foot of brick probably gives as good a house as can be built at reasonable cost; the pre-fabricated and metal houses were notoriously hot in summer and cold in winter; of all the roofing materials I have tested to see which gave the coolest house in summer and the warmest in winter, the old-fashioned thatch was by far the best. Next to it came green tiles, then all the usual types of tiles and slates, and at the other end of the list came corrugated iron, roofing felt and flat lead roof. But of course the results were not all comparable, since the houses were not all of the same pattern, or size, or built of the same material, or with the same pitch of roof. Yet I believe the thatch with its enormous volume of trapped air, and the flat lead roof would be first and last in any test of roofing materials.

The heat transmitted through windows by the solar radiation is very considerable and much greater than that through the walls; as the heat re-radiated by the interior of the room cannot pass out again through the glass a modified 'greenhouse effect' is noticeable in a south-facing room when the sun shines. In winter the glass readily takes up the temperature of the outer air, and the heat of the room near the window is quickly conducted through the glass and lost, so that flowery frost patterns on the inside of our windows are a common phenomenon on frosty mornings.

In winter the heat carried away when the warmed air we have made leaves the building by open doors and windows is so very considerable, that it is important to minimize the flow of air through the house to the amount necessary for proper ventilation. People so often throw up windows to air a room and then go away and forget them, until all the hard-earned warmth has been blown into the garden and the room for the time being is too cold to work in. They should remember that authorities often say that most buildings in Britain are over-ventilated in winter and have an unnecessary amount of air passing through them, so that most parts suffer from draughts especially when the wind is high outside. The cooling effect of a draught is proportional to the square root of the air speed, a draught of sixteen feet per second is twice as cooling as one of four feet per second. This is largely due to the air penetrating the clothing and distributing that layer of trapped air that exists there, and which is so efficient a non-conductor that it prevents excessive loss of body heat so long as it is still, but ceases to be efficient when continually replaced.

From this the importance of proper clothing in the winter is very apparent. It is not the amount and weight of the clothes we wear that is important, but the amount of trapped air we can enclose between the outer layers of our clothes and our skin, and this depends on the processing of the material more than on the material itself. For instance, such widely differing materials as wool, hair, cotton, slag, glass and steel, all have much the same heat insulating effect when processed in a fluffy form like cotton wool. Of these perhaps the most surprising is steel wool which is an efficient heat insulator while solid steel is an excellent heat conductor.

We can get some idea of the difference we should feel dressed in tightly fitting and in loose woollen clothes, by dressing-up one of a pair of matched thermometers in a series of pieces of wool cloth wrapped tightly and secured with a rubber band, and the other in the same number of pieces of wool cloth lightly wrapped and lightly tied. The range of temperature in a north-facing room in summer for the tightly packed cloth was about double that for the lightly packed. Similar experiments with heavy and cellular blankets on a bed showed that the cellular blankets were better heat insulators than the same number of thick blankets of greater weight.

Since it is important to keep any room near a temperature of 60° to 65° both in summer and winter if it is to be comfortable to work in, allowing of course that the user is likely to be wearing more clothing in the winter, it seems that the range of temperature of the room during every month is more important than the average temperature and this has been used in my experiments in my own house. My walls and roof appear to allow a quarter of the range of air temperature, taken close to the outside wall, to penetrate indoors. In an unfired north-facing room on a normally sunny day in summer the room has a range of temperature of 60° to 65° while the outside temperature ranges from 50° to 70°, and the room is fit to work in without any interior heating from about the end of May to the middle of September. A south-facing room, with no blinds or curtains drawn and normal ventilation, ranges from 61° to 70° while the outside temperature lies between 50° and 86°, and the room is fit to work in without heating from about the beginning of May to the end of September. But it must be remembered that the indoor minima and maxima lag about three hours behind the outdoor temperatures; this does not matter in the summer, but in the winter it means that a north-facing room will be at its coldest about 10 a.m. and at its warmest in the late afternoon.

Various formulae have been suggested for calculating the amount of heat required to keep a room a given number of degrees above the outside air; as these are generally based on the conductivity of the outer walls and take no account of windows, inside walls, ceilings and floors, they can at best be only first approximations. What are really required are definite amounts of gas or electricity which have been necessary to raise and keep a room at some specified temperature when the outside air temperature is also known. If full details are available of thickness and material of all the walls, size of the room and of window, and whether the house is a bungalow or house of several stories, with rooms above or below the room in question, a prospective buyer could assess his heating costs before buying with some degree of accuracy.

In my own bungalow a small north-facing room, 15 ft. long, 11 ft broad and 10 ft high, with two brick outer walls 12 in. thick and two inner walls 6 in. thick, ceiling with roof 10 ft. above and floor with foundations below, window 5 ft. by 5 ft., requires one therm of gas, or sixteen units of electricity, in a twelve-hour day to keep its temperature at 60° when the outside air temperature remains at 32° all day. This costs me 1s. 2d. in electricity, or about the same in gas per day, and rather less if the outside temperature rises a little during the day. My hall, running the whole length of the bungalow, is 40 ft. long, 6 ft. wide and 10 ft. high and has no outside walls but a glass door at each end leading to the south and north garden. With an electric thermovent heater, with the thermostat set at 60°, I can keep this hall warm all day in really cold weather in the winter by using twenty units of electricity costing me 1s. 6d. a day, and as all the rooms lead off this hall they too gain some heat from it when the doors are open.

On a winter morning with the thermometer standing at 25° outside and a thick fog, we may find our room after its morning airing has a temperature of 40° and a humidity of 90 per cent. At this temperature and humidity the water content of the air is low, and when we warm up the room to 65° we shall find that the humidity drops to 38 per cent. To raise this to 60 per cent we shall have to make use of the old-fashioned but efficient method of putting a bowl of water in front of the gas or electric fire until sufficient water has been evaporated into the room to raise the humidity. I have chosen the figure of 60 per cent humidity as desirable because what has been claimed to be the best relative humidity for efficient brainwork, and the results of an experiment at Fettes College, Edinburgh, years ago on a mathematical class showed that much better work was done on days when the relative humidity lay between 50 and 65 per cent than on days when it lay between 65 and 80 per cent. Certainly there is an urge to be up and doing in the stimulating north-east winds of spring with a low relative humidity. It is not for nothing that Scotland should be famous for oats and sheep; porridge to supply our bodily heat and wool next to our skin to conserve it are essential accompaniments of our Scottish climate.

Index

Aconite, 39
Aitken, Dr. John, 89
Ångström, Anders, 47
Antirrhinum, 94, 95
Ants, 84
Apple, 105
Apricot, 81, 82
Archangel, 105
Aspidistra, 107

Badgers, 19, 20
Beetles, 84
'Biology of Danish Spiders', 85
Bindweed, 39
Blackbirds, 12, 23, 24
Blow flies, 93
Bluebell, 58, 59
Burger, H., 59

Celandine, 39, 58, 59
Chase, J. L. H., 112
Chickweed, 87
Chlorophyll, 102, 107
Christmas rose, 73, 74
Chrysanthemum, 108, 109
'Climate and the British Scene', 81
Cloches
   condensation on, 115-17
   glass and plastic, 114-17
   light values, 115-17
   modern, 112
   old French, 111
   frost protection, 113
   temperature under, 113-16, 119

'Cloche Gardening', 112
Coltsfoot, 39, 47, 48, 61, 62
Cosmos, 109
Cranberry, 27
Cream cheese, 42
Crickets, 84
Cuckoo pint, 58

Dahlia, 13, 68, 109, 110
Daisy, 87
Daphne, 94
Delphinium, 94, 106
Dew, amount of, 90-2
   collected on plastic, 91, 92
   ponds, 90, 91
   theory of, 89
'Dissertation on the Growth of, Wine in England', 81
Dog rose, 39
Dog's mercury, 58, 59
Dormouse, 20, 21

'Essay on Dew', 89
Eser, Carl, 59
Evaporation, 27, 39, 45, 50-6, 84, 87, 88, 95, 96, 98
Evening primrose, 108
Everlasting flower, 87

'February fill-dyke', 65, 66
Fig, 82
Fox, 19
Frames, cold, 117, 119
   electrically heated, 117, 118
   temperatures in, 117, 119

INDEX

Frost
 damage, 27, 73, 74
 depth of penetration, 65, 76
 effect on hibernation, 21-3
 forecasting, 47, 68-73
 hollows, 68
 protection, 29, 67, 68, 74, 75, 113, 128

Geese, 121
Geranium, 109
Glass, 111, 114-17
Goat's beard, 108
Grant, Sir A., 97
Grapes, 14, 81, 82
Grass, 68, 78, 79, 86, 127
Grasshoppers, 84

Hall, Sir A. D., 98
Hedgehogs, 21-3
Hollyhock, 108
Home heating
 clothing, 132
 cooling effect of draughts, 131, 132
 comfortable humidity, 134
 comfortable temperature, 130
 cost of, 133, 134
 insulation, 129
 losses, walls and roof, 130, 131
 purpose of, 129, 130
Horse chestnut, 39
Humidity, 47, 49, 86-9, 95, 96, 99, 100
Hyams, E., 82
Hydrangea, 74

Irrigation, 57

Jasmine, 109

Keen, Dr. B. A., 57

Latent heat, 47, 73-5, 88

Laurel, 96, 103, 105
Lettuce, 109
Light values, 105, 115-17
Lilac, 109
Locusts, 13
Love-in-a-Mist, 109
Lupin, 94, 106

Manley, G., 81
Moles, 12, 42, 61
Monastic gardens, 14, 80, 81
Moulds and mildews, 88, 89
Mulches, 39, 45, 46, 59. 60, 62

Nectarine, 81
Nörgaard, E., 85

On Dew, 89
Orange, 105
Orchards, 42, 43, 67, 68

Partridges, 13
Pastures, fattening, 127, 128
 grown in rows, 122, 125, 126
 grazing of, 121-3
 herb strips, 123, 124
 new leys, 126, 127
 nurse crops, 124, 127
 palatability of, 122-4
 temperature, 128
Peaches, 81
Phlox, 109
Photosynthesis, 102-7
Pimpernel, 87
Plastics, 92, 114-17
Plover, 13
Plum, 81
Potatoes, 53-5, 75
Primrose, 58
Privet, 105
Pyrethrum, 114

Rabbits, 15-18, 82, 120

INDEX

Radiation, 48, 77, 78, 84, 102-6, 115
Rain
  capillary rise of, 39, 57
  evaporation of, 50-2, 56
  percolation of, 56, 57, 59
  transpiration of, 53-5
Ribwort plantain, 123
Rose, 36, 86, 94, 95

Snow, 62-5
Soils
  bacteria, 34, 35, 39, 50
  clay, 35, 36, 38, 40, 41, 57
  colour of, 37
  conductivity of, 37, 38, 45, 59, 60
  early and late, 38, 39
  loam, 35, 38, 40, 45, 56
  moisture, 35, 39, 55, 56, 59
  percolation, 56, 57, 59
  sand, 28, 38, 40, 56
Spiders, 85
Strawberries, 111, 112
Sunflower, 84, 105, 106-9
Sweet William, 110

Temperatures
  above crops, 82, 83
  air, 43, 63, 67, 77-84
  grass minimum, 78, 79
  growing, 39, 48, 49
  hibernating, 21-3
  in badgers' sets, 20
  in blackbirds' roosts, 23
  in cranberry marshes, 27, 31, 32
  indoors, 130-3
  in frost hollows, 68
  in orchards, 42, 43, 67
  in rabbits' burrows, 17, 18
  in windbreak zone, 98-100
  mean, 47, 48
  of different-coloured soils, 37
  of leaves, 53, 103-5
  optimum for bacteria, 35, 46, 47
  over tarmac, 79
  reduced by evaporation, 39, 88, 96
  soil, 17, 29-32, 37, 38, 43, 46, 58, 60, 88
  surface mulch, 45, 46
  under cloches and frames, 113, 115, 116, 119
  underground, 17, 43, 46, 58-64
  under mulches, 60
  under snow and frozen soil, 62-4
Thermometers, 44
The Soil, 98
Timothy, 53, 54, 100, 123, 125-7
Toad flax, 108
Tomatoes, 109
Transpiration ratios, 53-5

Viburnum, 94
Vineyards in England, 82
Vispré, F. X., 81

Walled gardens, 80, 81
Walls—South-facing, 79-81
Watercress, 84
Weeds, 120
Wells, Dr. W. C., 89
Windbreaks, 95-101
Wind damage, 94-6, 100
William of Malmesbury, 81
Williams, Stella S., 100
Wilson, Sir D., 122
Wood anemone, 58, 59, 105
Woodcock, Elizabeth, 64

Young, Arthur, 122